U0192352

Traditional Chinese Architecture Surveying and Mapping Series: Religious Architecture

ARCHITECTURE COMPLEX OF GAOPING

Compiled by School of Architecture, Tsinghua University
Edited by LIU Chang, WANG Guixiang, LIAO Huinong

China Architecture & Building Press

国家出版基金项目
NATIONAL PUBLICATION FOUNDATION

『十二五』国家重点图书出版规划项目

中国古建筑测绘大系·宗教建筑

高平建筑群

清华大学建筑学院 编写
刘畅 王贵祥 廖慧农 主编

中国建筑工业出版社

Contents

目录

Introduction

Since its inception in 1946, the School of Architecture at Tsinghua University has been committed to surveying and mapping traditional Chinese buildings, following the practice of the Society for the Study of Chinese Architecture (*Zhongguo Yingzao Xueshe*) that LIANG Sicheng, a driving force in the Society and founder of Tsinghua's architecture department (known as the School of Architecture since 1988), and his assistant MO Zongjiang brought with them to Tsinghua. Between 1930 and 1945, with members of the Society, LIANG visited over two thousand Chinese sites located in more than two hundred counties and fifteen provinces, and discovered, identified and mapped over two hundred groups of traditional buildings, including the famous Tang-period east hall (dating to 857) of Foguang Monastery at Mount Wutai—which was not an easy task because of the harsh working conditions in the secluded and relatively inaccessible villages in the countryside. In that same spirit, despite the difficult political circumstances from the 1950s through the 1970s, the School of Architecture conducted a systematic survey of historical buildings in the New Summer Palace (Yiheyuan). At the beginning of the Cultural Revolution in the late 1970s, all members of the faculty focusing on the history of architecture went to Hebei province under the leadership of MO Zongjiang to measure and draw the main hall of Geyuan Monastery in Laiyuan, an important Liao-period relic hidden in the remote mountains. This was followed by in-depth research and analysis. At the same time, those professors that specialized in Chinese architectural history (MO Zongjiang, XU Bo'an, LOU Qingxi, ZHANG Jingxian, and GUO Daiheng) led a group of graduate students to Zhengding in Hebei province, where they conducted component analysis and research of Moni Hall at Longxing Monastery, a Northern-Song timber-frame structure that had partially collapsed but was then in the process of being rebuilt. They also investigated nearby

导言

因为前辈学者梁思成及其助手莫宗江两位先生从中国营造学社继承的传统，清华大学建筑学院自创立以来，一直十分注重古代建筑实例的实地考察与测绘。尽管在 20 世纪 50 至 70 年代受到各种因素的影响与冲击，那时的清华大学建筑系，还坚持了对颐和园内一批古代建筑实例的系统测绘。改革开放刚刚开始的 1970 年代末，清华大学建筑历史方向的全体教师，就在莫宗江先生带领下，共同远赴偏僻的河北山区，考察测绘了创建于辽代的涞源阁院寺大殿，并对这座辽代木构建筑进行了系统研究。同是在那一时期，建筑历史教研室的莫宗江、徐伯安、楼庆西、张静娴、郭黛姮等教师，带领研究生赴河北正定，除了对正在落架重修的北宋木构大殿隆兴寺摩尼殿的大木构件进行现场分析研究外，还对正定及周边的古建筑进行了系统考察与调研。这种由老先生带队，

historical buildings in and around Zhengding. This practice of teamwork—senior researchers, instructors, and (graduate) students participating in the investigation and mapping of traditional Chinese architecture side by side—became an academic tradition at the School of Architecture of Tsinghua University.

Since the 1980s, fieldwork has been a crucial part of undergraduate education at the School, and focus and quality of teaching has constantly improved over the past decades. In the 1990s, professors like CHEN Zhihua and LOU Qingxi carried out surveying and mapping in advance of (re)construction or land development on sites all across China that were endangered. Since the turn of the twenty-first century, the two-fold approach—attaching equal importance to practice (fieldwork) and theory (teaching)—was widened and deepened. Sites were deliberately chosen to maximize educational outcome, resulting in a broader geographical scope and spectrum of building types. In addition to expanding on the idea of vernacular architecture, special attention was paid to local (government-sponsored) construction of palaces, tombs, and temples built in the official style (*guangshi*) or on a large scale (*dashi*), and to modern architecture dating to the period between 1840 and 1949. Students and staff have accumulated a lot of experience and created high-quality drawings through this fieldwork.

In retrospect, we have completed surveys of several hundred monuments and sites built in the official dynastic styles of the Song(Jin), Yuan, Ming and Qing all across the country. Fieldwork was always combined with teaching. Among the architecture surveyed are the (single- and multi-story) buildings in front and on the sides of the Hall of Supreme Harmony in the Forbidden City in Beijing; the architecture at Changling, the mausoleum of emperor Jiaqing located at the Western Qing tombs in Yi county, Hebei province; the monasteries on Mount Wutai, Shanxi province, including Xiantongsi, Tayuansi, Luobingsi, Pusading, Nanshansi (Youguosi), and Longquansi; Zhongyue Temple, Songyang Academy, and Shaolin Monastery in Dengfeng, Henan province; Xiyue Temple, Yuquan Court, and the Taoist architecture on the peaks of Mount Hua in Weinan, Shaanxi province; Chongan Monastery, Nanjixiang Monastery, Jade Emperor Temple (Yuhuangmiao) in Shizhang, and the temples of the Two Transcendents (Erxianmiao) in Xiaohuiling and Nanshentou, all situated in Lingchuan county of Shanxi province; and the upper and lower Guangsheng monasteries and the Water God's Temple in Hongdong, Shanxi province. In recent years, we have developed a specialized interest in the study of religious architecture of Shanxi province and investigated almost a dozen privately- or government-sponsored Song and Jin sites

教师与研究生集体参与，对古代建筑进行深入考察与测绘研究的做法，在清华大学形成了一个良好的学术传统。

1980年代以来，清华大学建筑学院始终在本科教学环节中，坚持讲授古代建筑测绘这门经典课程。这一传统在21世纪初的这十几年中始终延续。如果说，20世纪90年代由陈志华、楼庆西等教授带领的测绘教学，将相当的注意力放在了分布于全国多个省、市、自治区大量传统乡土村落建筑的抢救性测绘上，进入21世纪以来，清华大学建筑学院开展的这种结合本科教学的古建筑测绘教学与实践，覆盖的地域范围与建筑类型范围更为宽广：除了进一步拓展乡土建筑的测绘之外，在对各地留存的历代官式或大式建筑，如宫殿、陵寝、寺庙等建筑的测绘以及近代建筑的测绘上，也积累了大量测绘经验、图纸及丰富的调研资料。以古代官式建筑测绘为例，结合本科教学，我们先后完成了北京故宫太和殿前及两侧门殿、楼阁与朝房建筑，河北易县清西陵昌陵完整建筑群，山西五台山显通寺、塔院寺、罗睺寺、菩萨顶、南山寺（佑国寺）、龙泉寺等多座整组寺院建筑群，河南登封中岳庙、嵩阳书院、少林寺古建筑群，陕西渭南华山西岳庙、玉泉院及华山山顶各道观古建筑群，山西陵川崇安寺、南吉祥寺、小会岭二仙庙、南神头二仙庙、石掌玉皇庙，以及山西洪洞广胜上寺、广胜下寺、水神庙等数百座古建筑实例的测绘，其时代的范围覆盖了宋（金）、元、明、清等历代木构建筑遗存实例。近几年，我们又将测绘的重点放在了高平、晋城等晋中及晋东南地区，

located in central Shanxi (Jinzhong) and southeastern Shanxi (Jindongnan), specifically in Gaoping and Jincheng counties. This includes the Youxian, Chongming, and Kaihua monasteries and the Two Transcendents Temple in Xilimen. Additionally, supported by the State Administration of Cultural Relics, the head of the Architecture History and Historic Preservation Research Institute at the School of Architecture, Liu Chang, led a group of students to map and draw the main hall of Zhenguo Monastery in Pingyao, a rare example from the Five Dynasties period. The survey results have been published. Tsinghua fieldwork in Shanxi has become an annual event that is jointly organized almost every summer by the faculty of the School of Architecture, including professors engaged in research on non-Chinese architecture, in cooperation with their graduate students.

It is worth mentioning that since 2007, the School has worked in collaboration with the well-known company China Resources Snow Breweries Ltd., which supports the transmission and dissemination of knowledge on traditional Chinese architecture and provides funds for the School's research and field investigation activities. Drawing on the support from industry allowed us greater initiative and flexibility, and we were thus able to carry out research on and survey often overlooked but no less important Song-Jin monuments in central and southeastern Shanxi.

Our years-long fieldwork has not only enabled us to teach students subject knowledge about scale, material, form, and decoration of traditional Chinese architecture as well as a sense of appreciation for the old, but has also provided us with plenty of data for monument preservation practice and research. China Architecture and Building Press spared no effort in compiling and publishing the results of the fieldwork in 2012. Publication has also been supported by the National Publishing Fund. This highlights not only the importance of our contribution to architectural education at the national level but also shows its significance for the transmission, development, and revival of traditional Chinese architectural culture both at home and abroad. In order to expand the reach of this work to an international audience, *the Traditional Chinese Architecture Surveying and Mapping Series* is being published bilingually. Based on the past ten years of fieldwork, we have now compiled five volumes, namely *Mount Wutai's Buddhist Architecture* (Traditional architecture on Mount Wutai, Shanxi), *Architecture Complex of Songshan* (Traditional architecture in Dengfeng, Henan), *Mount Hua's Yuemiao and Taoist Temples* (Traditional architecture on Mount Hua, Shaanxi), *Architecture Complex of Hongtong* (Traditional architecture in Hongtong, Shanxi), and *Architecture Complex*

对包括高平游仙寺、崇明寺、开化寺、西李门二仙庙等在内的十余座宋金建筑群，进行了全面而系统的测绘。这一期间，在国家文物局的支持下，建筑历史与文物保护研究所刘畅老师还带领研究生对五代时期创建的平遥镇国寺大殿等建筑进行了精细测绘，并出版了测绘研究成果。此外，清华大学建筑学院的测绘工作，几乎每年都是由全体建筑历史教师共同合作，并带领研究生们共同完成的。从事外国建筑史教学的老师，也不例外。

特别值得一提的是，自2007年以来，清华大学建筑学院与国家知名企业华润雪花啤酒（中国）有限公司建立了良好的合作关系。该集团不仅支持中国古建筑知识的传承与普及工作，也对清华大学建筑学院中国古代建筑研究及古建筑测绘工作给予了直接的支持，使得我们的古建筑测绘工作变得更为主动和更具选择性。一大批珍贵的山西晋中及晋东南地区宋金时代建筑实例的测绘与研究，就是在这样一个前提下得以顺利开展与完成的。

坚持数十年的古建筑测绘工作，不仅在培养学生对传统中国建筑的尺度、材料、造型与细部装饰的认知与感觉上起到了直接的影响，而且也为各地文物建筑保护与研究工作，提供了相当充分的资料支持。

2012年，中国建筑工业出版社花大气力组织了汇集全国重点院校建筑系古建筑测绘成果的中国古代建筑测绘大系的编辑出版工作。这一工作也获得了国家出版基金的支持。这不仅是对高校建筑教育成果的一份支持，也是对中国传统建筑文化传承、发展与复兴的一份支持。正是在这样一个背景与前提下，我们对近十余年来考察测绘的古代建筑案例加以整理，分别编汇了包括《五台山佛教建筑》《嵩山建筑群》《华山岳庙与道观》《洪洞建筑群》《高平建筑群》5册古建筑测

of Gaoping (Traditional architecture in Gaoping, Shanxi). The architectural drawings presented in these books are carefully selected and screened by Tsinghua professors. They only show a part of our comprehensive surveying and mapping work, but still cover a whole spectrum of geographic regions and time periods. Thus, they contain information of high academic value that may serve as a reference for future study and for the protection of cultural heritage. It is hoped that our work will help to promote interest in and improve understanding of traditional Chinese architecture, not only among Tsinghua students (through hands-on experiences in the fieldwork) but also among architectural historians and professionals engaged in monument preservation at home and abroad.

As a final thought, let me shortly address the workflow. The drawings presented here are based on survey and working sketches drawn up on site during several years of fieldwork conducted by Tsinghua professors together with graduate and undergraduate students. Back home, the measured drawings were redrawn over months of diligent work by graduate students with computer-aided software to achieve dimensionally accurate and visually appealing results, a project that was completed under the supervision of LIU Chang, head of Tsinghua's Architecture History Institute, and the Tsinghua professors LIAO Huinong and WANG Nan, as well as TANG Henglu and his colleagues from the WANG Guixiang Studio. We would like to take this opportunity to thank the professors, students and colleagues who participated in the fieldwork and its revision.

Our final thanks go to LI Jing, assistant researcher at the Architecture History Institute here at Tsinghua. Next to participating in surveying and mapping, she organized the development of the book and moreover, made this book possible in the first place.

WANG Guixiang, LIU Chang, LIAO Huinong
Architecture History and Historic Preservation Research Institute, School of Architecture, Tsinghua University
December 5, 2017

Translated by Alexandra Harrer

绘图集，作为这套『中国古建筑测绘大系』的部分成果。尽管这只是我们多年测绘成果的一部分，但也是清华建筑历史学科教师们仔细筛选、认真校对、充分整理之后的较具典型性与参考性的成果。这些成果不仅地域覆盖面大，而且建筑遗存的时代跨度也相当长，具有十分重要的学术价值。希望这些成果对高校建筑系学生们学习古建筑，建筑历史学者研究古建筑，以及文物保护工作者从事文物古建筑的保护与修缮，能够起到积极的推动作用与重要的参考价值。

最后要提到的一点是，除了参与测绘的教师、研究生与本科生多年历尽辛苦的测量与绘图工作之外，此次清华大学建筑学院承担的这5册测绘图集，也经由建筑历史与文物保护研究所刘畅、廖慧农、王南和他们的研究生，以及王贵祥工作室团队的唐恒鲁等同仁们在既有测绘图纸基础上，经过数月认真仔细的线条分层、图面调整、数据校对、图面完善等缜密修复工作，在这里也要向参加测绘图整理的老师、同学和同事们表示感谢。

还应该特别提到的是建筑学院建筑历史与文物保护研究所的助理研究员李菁博士，她不仅参加了多次测绘，还为这套书最后的编辑与出版做了大量相关工作。这里一并表示感谢。

王贵祥、刘畅、廖慧农
清华大学建筑学院 建筑历史与文物保护研究所
2017年12月5日

Preface

Tsinghua University fieldwork in Gaoping actually began in 2010, when I first brought some graduate students to Shanxi province to learn on-site and explore directions for future surveying and mapping of historical architecture. Official teaching activities then started in 2015 as part of the "Large Surveying and Mapping Corps" of the School of Architecture and were supported by the local cultural relics bureau. In the following three years, we were able to experience numerous cultural monuments in Gaoping protected at the county, city and state levels. From the comprehensive data we collected over the years, we have selected five National Priority Protected Sites in Gaoping that we would like to present to the reader in this book—the Chongming, Youxian, Kaihua, and Zisheng monasteries and the Temple to the Two Transcendents (Erxianmiao) in Xilimen.

Chongming (Exalted Brilliance) Monastery

Chongming (Exalted Brilliance) Monastery is situated on the eastern slope of Shenfo Mountain fifteen kilometers southeast of the county-level city of Gaoping in Shanxi province. Before the retaining wall that blocks our view today was built in front of the monastery, the complex was surrounded by untrimmed fields interspersed with yellow loess soil. After a final bend of the road winding up the mountain, one was able to catch a spectacular glimpse of the otherwise rather modestly designed buildings in the mist of the early morning. Today, the architecture is still nestled into the gentle rolling hills and evokes a peaceful and quiet impression (Fig.1). Buildings are arranged in two small courtyards and comprise a front gate (*shanmen*), a middle Buddha Hall, and a back hall all aligned on the central axis and, on the left and right courtyard sides, drum and bell towers and east- and west-side buildings (*peidian*). Additionally, there are auxiliary buildings to both sides of the main halls they are facing.

One of the great challenges we faced was dating the approximate age of the historical buildings.

序言

山西高平与清华大学古建测绘课程的渊源始于2010年本人和研究生们为教学而做的踏勘。在当地文物局的大力支持下，自2015年始，清华大学建筑学院的『测绘大部队』正式逐步开始了在高平的教学活动。在2015–2017年这三年中，我们的测绘队伍亲眼观察、亲身体验、亲手抚摸过高平不少的国家级、省级、市县级文物保护单位，而这次汇集、筛选并呈现给读者的是五处国家重点文物保护单位——崇明寺、游仙寺、开化寺、资圣寺和西李门二仙庙，只是高平瑰宝的少数代表。

一、崇明寺

崇明寺，位于高平市城东南15公里的圣佛山东麓。在今天寺前大坝一样的挡

图1 崇明寺俯视 图片来源：赵波拍摄

图2 崇明寺中佛殿 图片来源：徐杨拍摄

图3 崇明寺淳化二年碑（991年建成） 图片来源：徐杨拍摄

Fig.1 Overlooking the Chongming Monastery
Source: Photographed by ZAHO Bo
Fig.2 The middle Buddha Hall of Chongming Monastery
Source: Photographed by XU Yang
Fig.3 The Stele of Chongming Monastery was built in the second year of Chunhua (A.D. 991) Source: Photographed by XU Yang

How do we know that the middle Buddha Hall was built during the Song dynasty (in 971 (Fig.2), the fourth year of the Kaibao reign period) but the rear hall in the Ming dynasty? Starting with a visual assessment, the middle Buddha Hall is a three-bay wide and six-rafter deep structure with a single-eave hip-gable roof covered with semi-circular tiles and decorated with glazed dragon-head ridge ornaments (*wenshou*). Column-top bracket sets are of seventh rank (out of eight possible ranks) with double-tier *huagong* (perpendicularly projecting bracket-arm) and double-tier *xia'ang* (descending cantilevers). Such bracketing is quite rare in North China, and resembles only that of another late tenth-century building, the roughly contemporary Wangfo (Ten-thousand Buddha) Hall (d. 963) at Zhenguo Monastery in Pingyao, located several hundred kilometers away. However, gathering and processing visual information on site to make a first assessment of a building is a crucial part of fieldwork but must be preceded by in-depth study of relevant historical documents. In a joint effort, Tsinghua teachers and graduate students found a surprisingly rich body of records that contain information about Chongming Monastery. Thus, we can trace back the history of the monastery to the Tianbao reign of the Northern Qi dynasty, based largely on the information from a local Qing-period (Daoguang reign) gazetteer (*Xiuwuxian zhi, juan* 10 includes *Jinshizhi*) that records the inscriptions of two temple steles—*Wuliyuan Chongmingsi bei* dating to 1350, the tenth year of the Zhizheng reign period, and *Chongxiu Wuliyuan Chongmingsi bei* dating to 1534, the thirteenth year of he Jiajing reign period. We also know about rebuilding carried out in the Northern Song dynasty from a stele (*Chuangxiu […] shenfoshan Chongmingsi ji)* dating to 991 (the second year of the Chunhua reign period) that has was preserved below the front eaves of the main hall. According to the stele inscription, reconstruction took more than twenty years, starting in the early Kaibao reign period (969—976) and lasting until 991 when the stele was erected (Fig.3). (The monastery thus must have been destroyed at some point before that.) Yet the next line of the stele text is puzzling, telling us not only about the collecting of building material but also about the far-reaching search for a master carpenter to construct the monastery buildings. But how far did the monks have to look? Where did they find a skilled artisan? Can it be that this monastery and Thousand-Buddha Hall at Zhenguosi are designed and built by the same person? These questions to which we devoted a great deal of attention still await further research.

Youxian Monastery

Youxian Monastery, formerly known as Cijiaoyuan or Court of Benevolent Teaching, is situated ten kilometers south of Gaoping on the slope of Youxian Mountain from which its name derives. A small road branching off from the provincial highway winds its way up toward the monastery through gently rolling hills. Although not high, the hills add

土墙修建之前，寺院完全被参差的绿色庄田和共间露出的斑驳黄土围绕。山回路转之际，在薄雾弥漫的清晨，最是激荡人的心胸（图1）。那只是一座由山门、中佛殿、后殿、东西钟鼓楼、配殿、两庑组成的不大的院子，静静地躲藏在山丘起伏的那一侧；怎知道院中的中佛殿当是北宋开宝四年（971年）所创（图2），后殿也是明代的建筑。这座中佛殿殿身面阔三间，进深六椽，单檐九脊顶，筒板布瓦盖顶，上施琉璃吻兽；柱上用双杪双下昂七铺作，是北方屈指可数的七铺作代表作，更与数百公里外的平遥镇国寺万佛殿的七铺作如出一辙。现场考察和测量工作是测绘的第一要务，而文献检索工作甚至必须是现场工作之前的第一步。清华测绘团队的史料工作由辅导教师和研究生承担。崇明寺历史文献给我们的启发也是异常深刻的。我们知道清道光《修武县志·金石志》中记载了两通古碑——至正十年（1350年）的《五里源崇明寺碑》和嘉靖十三年（1534年）的《重修五里源崇明寺碑》能够把建寺的历史推至北齐天宝年间；我们还知道，保存在大殿前檐的淳化二年《创修□□圣佛山崇明寺记》。令人倍感欣喜的是，淳化二年碑记中提到『当寺始自开宝之初』，即968—976年之前段，工程营造『历二十年余』至淳化二年成碑（图3）；而碑文之中最为令人浮想联翩的是碑文提到在寺院建造过程中，『取鹤栖之梁栋，远寻哲匠，结构斋堂』。远寻哲匠，远到哪里呢？莫非与成于963年的平遥镇国寺万佛殿有什么渊源呢？期待这些测绘教学所无法回答的问题能够在未来的研究中找到答案。

二、游仙寺

游仙寺，原名慈教院，在高平市城南10公里的游仙山。从省道拐入通往游仙寺的小路，是一段田野中缓缓起伏上升的盘旋路。山虽不高，寺院气势非凡。这是一座坐北朝南深三进，并带左右跨院的大型

图5　游仙寺毗卢殿　图片来源：姜铮拍摄

图4　游仙寺俯视　图片来源：李沁园拍摄

Fig.4 Overlooking the Youxian Monastery Source: Photographed by LI Qinyuan
Fig.5 The Pilu Hall of Youxian Monastery Source: Photographed by JIANG Zheng

momentum to the site. The complex is oriented southward and comprises three courtyards and, on the left and right sides, two groups of buildings that extend beyond the confines of the walled monastery enclosure. After entering through the front gate known as Chunqiu (Spring and Autumn) Building, there are aligned on the central axis Pilu (Vairocana) Hall, Sanfo (Three Buddha) Hall, and Qifo (Seven Buddha) Hall (Fig.4, Fig.5).

The history of the site can be traced back to the Northern Song period, although historical records are not consistent (and further research might place construction of the site to an earlier date). In a stele inscription from 1041 (the second year of the Kangding reign period of the Song dynasty), the "presented scholar" (Jinshi) and county official YAN Yizheng commemorates the founding of the monastery, arguing that the donor Zhang Gongming built a three-bay Buddha Hall here together with abbot Ji Yi, who had went to Yantang Monastery in Shangdang (present-day Gaoping) region to learn; "neither seeking fame nor wealth, [only following] his own karma", Ji Yi was "an abbot who excavated soil", a line that suggests his building activity at Youxian Monastery. Another stele (*Shifang Cijiaoyuan shidiji*) from 1304 (the eight year of the Dade reign period of the Yuan dynasty) gives more hints about construction and destruction of the site, saying: "from the begin of the construction until today, a period of more than two-hundred years has passed, during which the site was plundered twice by soldiers but still not abandoned". A stele inscription from 1514 (the ninth year of the Zhengde reign period of the Ming dynasty) by the Chan master Dabao specifies the date but suggests another planner and developer, saying: "from historical steles we can know that construction started in the first year of the Chunhua reign period of the Great Song, when monk Hui Gong was abbot; at that time the county official JI Zizheng (with the title Quannongguan), who was a devout Buddhist and admired the monks, assisted in collecting material and constructing the monastery, and bestowed a plaque inscribed with the name Youxian Monastery." Finally in 1752 (the seventeenth year of the reign of the Qing emperor Qinglong), county-school scholar XU Nakui argued that Youxiansi was allegedly erected in the Dali reign period of the Tang dynasty.

Today it is commonly agreed that Pilu Hall was built in 990 (the first year of the Chunhua reign period of the Northern Song dynasty). Although exact dating of the historical buildings requires in-depth research, experienced historians may "sense" the styles of different dynasties through visual assessment of the architecture on site. For example, grey bricks used for floor tiling and wall construction create a certain historical atmosphere at Youxian Monastery, which is why we can conclude that its architecture—except for the recently renovated masonry buildings—dates no earlier than the Ming period. At that time, bricks were the most common building material, similar in popularity to reinforced concrete today. But eaves, windows and doors speak a different language—

建筑群。中轴线上，山门『春秋楼』之内，大殿分别被冠以毗卢殿、三佛殿、七佛殿之名（图4、图5）。

寺内碑刻中记载寺院的历史可以追溯至北宋。不过古人的记述也不一致，留给今天文人回味的空间：宋代康定二年（1041年）乡贡进士闫仪正记载了『河西檀那九河张公明』，后人为纪念张公与『上党延唐寺受业释继一』『同构成佛殿一所三间』的往事，再之前，继一和尚只是『拨土住持』，『淡泊随缘』；元代大德八年（1304年）《十方慈教院施地记》说，『自始迄今，二百余年，凡两遭兵劫，曾不废坏』；明代正德九年（1514年）大宝禅师写道，『古碑有云，始于大宋淳化元年，有僧辉公讲主□□于此时，本邑有劝农官姬子政，仰师道德，督众鸿材，创成寺宇，蒙敕赐额，曰游仙寺』；清代乾隆十七年（1752年），邑庠生许纳揆说『游仙寺建自唐大历年间』。

前殿毗卢殿的始建年代被学者推断为北宋淳化元年（990年）。尽管严谨的古建筑断代需要大量研究工作，而阅历深厚之人能够体会不同朝代的气息。游仙寺院落之中的气息更多的是青砖墁地、青砖墙面的气息，除了那些近年修缮砌筑的以外，古老者应当也不早于明代建城运动——那时候，曾经如同今日轻信钢筋混凝土一样热衷于用砖；而那些从檐下、从门窗静静溢漫开来的，是木头的气息，是手艺的气息，是不同年代建筑设计者和使用者的气息。一周现场工作之后，学生们和我们问起最多的，还是体会之外的点滴心得。

图6 开化寺俯视 图片来源：赵波拍摄

Fig.6 Overlooking the Kaihua Monastery Source: Photographed by ZAHO Bo

that of sophisticated timber framing and craftsmanship. Even after the one-week survey, we were still surprised and delighted by the artistry we found on site.

Kaihua Monastery

Kaihua Monastery is located in Wang village, Chen township, Gaoping prefecture, Shanxi province. From its elevated position on the southern flank of Sheli Mountain, it overlooks the worshippers who come to visit the site. Entering though a monumental front gate (*shanmen* also known as Dabei [Great Pity] Pavilion), the complex consists of a main courtyard and a parallel courtyard attached to its side. The ritual axis of the main courtyard comprises, in addition to the front gate, the Treasure Hall of the Great Hero (Daxiongbaodian) and Yanfa (Discoursing-on-the-Law) Hall (destroyed and reconstructed). Organized in a symmetrical manner on the sides of the axis are drum and bell towers, east- and west-side buildings (*wufang*), Sandashi (Three Great Beings) Hall and Dizang (Kṣitigarbha) Hall, Wenchang dijun (God of Culture and Literature) Pavilion and Shengxian (Sages) Pavilion, and Guanyin (Avalokiteśvara) Pavilion and the Pure Quarters of Vimalakīrti (Weimo jingshi). The side courtyard was the living and working space of the abbot and the monks (Fig.6).

Although the two-story Dabei Pavilion is the tallest building of the monastery, the principal building is the Treasure Hall of the Great Hero, a timber-frame structure with unique decorative and narrative polychrome wall murals (*bihua*) and architectural components (*caihua*) that have survived in their current form from the Song. By contrast, Guanyin Pavilion is associated with the succeeding Jin dynasty. An inscription carved into one of the front eaves columns dates the pavilion to 1141 (the first year of the Gaiyuan reign period of the Jin dynasty; namely the twelfth month of the Xinyou year of the sexagenary cycle). The pavilion construction is in line with the date and shows features typical for that period: a large eaves architrave (*yan'e*) spans across the three front bays; the central-bay columns are shifted to the sides by almost half a bay; supporting timbers (*chuomufang*) for additional abutment of the architrave are installed in the shortened side bays; tapering to a rounded tip (shaped like a Cicada belly), their heads penetrate the column shafts and reach into the enlarged central bay.

The fieldwork focused on the Northern-Song Treasure Hall of the Great Hero. The biggest challenge was the documentation of the decorative painting of architectural components (*caihua*) and of the narrative painting on the walls (*bihua*). The murals on the eastern wall are damaged and subsequently restored or altered by later generations; the murals on the western wall are comparatively complete, and depict scenes from the Bao'en Sutra (or Great Skillful Means Sutra on the Buddha's Repayment of Kindness) . An inscription was

三、开化寺

开化寺位于山西省高平市陈区镇王村，依舍利山而建俯瞰前来朝拜的众生。寺院是雄伟山门引领的一套主院和一进旁院。主院中轴线有山门（大悲阁）、大雄宝殿、演法堂（原作已毁），左右依次由钟鼓楼、东西庑房、三大士殿和地藏殿、文昌帝君阁和圣贤阁、观音阁和维摩净室；旁院则为方丈院（图6）。

建筑群之中，大悲阁为二层楼阁，最为高大；大雄宝殿是当之无愧的主角，木结构、建筑彩画、室内壁画都基本保留了宋代的原作，堪称现存地面建筑中绝无仅有的案例；观音阁则被专家认定为金代建筑，尤以柱上施三间之阔的大檐额，同时明间柱向两侧移开，次间下施绰幕枋，至明间出蝉肚形式为代表特点，前檐柱上刻有题记，曰『皇统改元元年（金，1141年）岁次，辛酉，十二月』。

学生们测绘的重点是北宋的大木结构，但是最为挑战的是需要每人都不得触碰殿内的壁画和彩画。这里的壁画，东壁部分画面已经漫漶不清，部分经过后人补绘；西壁相对完整，为报恩经变。考察画上『丙子六月十五粉，此西壁画匠郭发记并照壁』的题记，此丙子年是1096年，对照《泽州舍利山开化寺修功德记》中的『始以元祐壬申正月初吉绘修佛殿功德，迄于绍圣丙子（1096年）重九，灿然功已』一段文字，二者相校，脉络清晰。这里的彩画，《泽州舍利山开化寺修功德记》上另外一段文字：『夫

found on the west wall that gives the date of the fifteenth day of the sixth month of 1096 and mentions the name of the local artist Guo Fa. Together with a line from a temple stele (*Zhezhou Shelishan Kaihuasi xiugongde ji*) saying "the merit of painting and repair of the Buddha Hall began on Chuji day in the first month of the Renshen year of the Yuanyou reign period, and lasted to the ninth day of the ninth month of 1096 (Double-nine Festival of the Bingzi year of the Zhaosheng reign period), the results were brilliant and bright", we can thus know that the wall-painting program was completed by the end of the eleventh century. This suggests that the hall had been ready then for such decoration and must have been built at some point before 1096. Another line of the same inscription compares the polychrome painting of the hall to that of the homes of princes, dukes, and high-rank officials, praising its beauty and craftsmanship. But polychrome painting is like clothing: it should not only be aesthetically appealing but also protect against wind and cold. Yet on the outside, the original decoration that once protected the wooden construction below the eaves has long been lost due to the weather and replaced by a simple terracotta-red wash. Only the exquisitely painted space at the interior is still relatively intact. We were all deeply moved when we saw the beauty of the decorative painting that had lasted for centuries.

Zisheng Monastery

Zisheng Monastery is one of the many historical monuments still extant in Dazhou village. The west village gate has also survived—Zhenwu (True Warrior) Pavilion atop the passageway on the ground floor gives the gate a dynamic appearance full of vigor. There are a large number of well-preserved traditional houses in the village, and their good condition makes the survival of the monastery until present more understandable. Zisheng Monastery is located close to the entrance of the village and orientated southward. Another village gate known as Nanbao (Southern Barbican) Pavilion stood originally south of the monastery, but only the adjacent Guanyin Pavilion has survived. Although located slightly off the central monastery axis, Guanyin Pavilion still could have been part of the complex at some point. Today, the monastery consists of two courtyards. Aligned on the central axis are a front gate (*shanmen*), a middle hall (known as Pilu [Vairocana] Hall), and a rear hall (known as Leiyin [Sound of Thunder] Hall), all of which were repaired on several occasions in the past. On the sides of the central axis stand buildings that were built in recent years including bell and drum towers, east- and west-side buildings of one story (*peidian*) or multiple stories (*peilou*), and ear buildings (*duodian*) flanking the rear hall. The current layout differs from the condition of the site as orally transmitted by the local villagers from generation to generation.

There are seven stone steles currently stored in the monastery. Some steles were relocated from other sites in recent years. Two steles date to the Yuan period. Six steles have lengthy

释氏选梓人，资豫樟，以兴楼观；尚丹雘，绘文章，以雕□墉……其上与王公大人第舍相侔者，何谓也？」这不正说明当时的大殿彩画绚烂吗？只不过彩画如衣服，要好看，还要挡风保暖。于是大殿外檐已经遍布土红色的刷饰，只有内檐依然绚丽。顾盼之余，不禁由衷感慨学生之幸，我等之幸。

四、资圣寺

资圣寺所在的大周村古老而遗迹众多。村西门犹在，门洞上的真武阁气势不凡。村子中留有大量传统民居，是寺院赖以存在的基础。资圣寺就在村口不远，坐北朝南。原来寺南还曾有南堡阁作为南村门，它旁边的观音阁至今耸立。现存寺院建筑院落两进，中轴线上外有观音阁，前有山门，其后中殿，名曰毗卢殿，收于后殿，名曰雷音殿，均系历史建筑，历经修缮保护；两翼现状前有钟鼓楼，后续前后院东西配楼、配殿、后殿东西朵殿，均系地方近年新建，与乡民口述历史原貌存在较大差距。寺内碑刻，除了近年零星自他处迁来碑刻卧存院内之外，还存碑碣7通，其中元代造像碑2通，6通带有大量文字信息，还提到了唐末五代时期开化寺的立寺人大愚禅师和金元之交执掌资圣寺的紫岩钦禅师。回到建筑，寺院中轴线上，算上寺外的观音阁，毗卢殿可以一直追溯到北宋元丰五年，其他的则主要

inscriptions. The stele inscriptions mention the Chan master Dayu, who was the founder of the monastery in the late-Tang or Five Dynasties period, and the Chan master Ziyanqin, who was in charge of Zisheng Temple by the turn of Jin and Yuan dynasty. Nevertheless, the actual buildings are of more recent date. We can trace Guanyin Pavilion (although located outside the current complex) and Pilu Hall back to the fifth year of the Yuanfeng reign period of the Northern Song dynasty, but the other architecture probably dates to the Ming period. Although the Song timber-frame structure of Pilu Hall shows traces of later restoration, we still can discern a dramatic architectural dialogue that took place between Northern-Song large-scale monastery construction and the local style of a village temple (Fig.7).

Xilimen Erxian Temple

Xilimen village is located on the side of a public road. It is connected to the Two Transcendents' cult centering around two native sisters, which most likely started in the famous Cituan (Purple) Cave on Chirang Mountain in Lingchuan county, southeastern Shangdang region. The Temple of Erxian is hidden atop a small hill facing away from the village to the south. The temple consists of only one spacious courtyard. Outside in front of the complex, stands a stage building that does not match the temple layout. The front gate (*shanmen*) is a three-bay structure. It rises straight up from the ground and has brick walls that are interwoven with bracket sets and continued on both sides by an enclosing wall. The main hall with its large open front platform (*yuetai*) is spectacular and is encircled by east- and west-side buildings, corridors, and a worship building (*chonglou*). The rear hall is extremely simple and differs from the typical Erxian temple design where the rear hall acts as the main hall.

The construction of the main hall is neatly designed with side frameworks flanking the central bay that consist of three columns and a two-rafter beam (*rufu*) "abutting" a four-rafter beam (*sichuanfu*), thereby creating a (two-rafter) front gallery. Bracket sets are of fifth rank (*puzuo*), with one *huagong* (perpendicularly projecting bracket-arm) and one *ang* (cantilever), although neither their placement nor their design is consistent in itself. The central front bay has one intercolumnar bracket set. *Tiaowo* (ascending cantilever-like timbers) are installed at the interior projection of a bracket set to carry the load of the lowest roof purlin (*xiapingtuan*) above. The polychrome paintings on beams, joists and bracketing members inside the hall are well preserved, and the walls also show traces of color, but the decorative program of the hall awaits further research. The stone doorframe is exquisitely carved with forceful lines, and lively beasts embellish the threshold. More importantly, the stone lintel is inscribed with the date 1157 (the second year of the Zhenglong reign period of the Jin dynasty), and the inscription also tells us that the villagers were together responsible for the erection of the gate (Fig.8, Fig.9).

成于明代。端详大殿的木结构，后代修缮痕迹的掩映之下，依然可见一幅北宋年间名山大寺和乡村寺院之间交流的图景（图7）。

五、西李门二仙庙

西李门村就在公路边。或许是和二仙奶奶『显圣迹于上党郡之东南陵川县之界北地号赤壤山名紫团洞』有关，二仙庙藏在村子南侧的小山岭上，背向村庄。这是一座简单而宽敞的院落，院前保存着古代戏台的基座和无法与之相配的建筑。山门三间耸立，斗栱交织，两翼砖墙围护。院内当中的大殿最为壮观，前有大型月台，又有东西配殿、廊庑、崇楼为之衬托；后殿反而非常简朴，与很多二仙庙后殿为主的院落布局并不相同。

大殿的结构非常严整，方三间，当心间东西间缝为乳栿对四椽栿用三柱，留出前廊；斗栱则采用五铺作，杪昂搭配，不拘一格，主要布置在柱头和转角，仅于正面当心间用一朵补间，屋内挑斡上彻下平槫；殿内尚存比较完整的梁枋、斗栱彩画，壁面也保留不少色彩痕迹，值得深入研究；大殿石门框用材做工细腻，线脚挺拔，门枕上蹲兽活泼，更加重要的是，石门楣上清晰可见金正隆二年（1157年）的题记『晋城县莒山乡司徒村众社民施门一合』（图8、图9）。

Fig.7 The Daxiong baodian of Zisheng Monastery Source: Photographed by JIANG Zheng
Fig.8 The main hall of Xilimen Erxian Temple Source: Photographed by JIANG Zheng
Fig.9 The Inscription on the Stone Lintel of Xilimen Erxian Temple Source: Photographed by JIANG Zheng

图7 资圣寺大雄宝殿 图片来源：姜铮拍摄

图8 西李门二仙庙大殿 图片来源：姜铮拍摄

图9 西李门二仙庙石门楣题记 图片来源：姜铮拍摄

The platform in front of the main hall uses grey bricks and sandstone and takes the form of a podium inside a Buddhist shrine shaped like Mount Sumeru (*xumizuo*). The sandstone has eroded, but the outstanding and rich sculptural program is still visible. Of particular note, the waist section of the three-part platform is decorated with line engravings of actors and dancers as well as with dragons in high-relief. In between are small sandstone columns completely packed with lions, tigers, and deer. Such relief pillars are known in *Yingzao fashi* as *geshen banzhu*. Furthermore, an offering hall (*xiandian*) once stood atop the platform, according to the inscriptions of two stone steles standing in front of the rear hall from 1158 (the third year of the Zhenglong reign period of the Jin dynasty) and 1163 (the third year of the Dading reign period). Apart from expressing regret at the financial constraints and cursoriness of later stonemasons, it is fortunate that the exposure to rain through the passing years was far more damaging to the steles than the atrocities of succeeding generations of the morally inadequate.

The students' fieldwork posed many exciting questions that still await research in the future. After my brief introduction and retrospective, as the editor of this volume, I would like to express my sincere thanks to the leaders and members of the surveying and mapping teams—especially to the professors of the Architecture History and Historic Preservation Research Institute at Tsinghua University, namely HE Congrong (Chongming Monastery), Jia Jun (Youxian Monastery), LI Luke (Kaihua Monastery) and WANG Guixiang (Xilimen Erxian Temple), and to their doctoral and postdoctoral students who greatly contributed to the success of our fieldwork. I owe special gratitude to those students (different graduation years) who provided support in terms of teaching, research, and counseling over the years-long fieldwork, and I would like to list their names here: XIN Huiyuan (from South Korea), HUANG Wenhao (from South Korea), PU Zhaoyan (from South Korea), YANG Shu, ZHANG Yichi, WENG Fan, LI Qinyuan, SUN Lei, XU Teng, ZHAO Sarina, FU Na, WANG Xichen, HE Wenxuan, LIU Mengyu, JIANG Zheng, ZHAO Shoutang, XU Yang, JIANG Zhe, LIU Xiaoxiao, and ZHAO Bo.

LIU Chang
Architecture History and Historic Preservation Research Institute,
School of Architecture, Tsinghua University

Translated by Alexandra Harrer

大殿正前的月台采用了青石、砂岩混用的须弥座形式。砂岩虽已剥蚀，仍能看出雕刻富丽，不同凡响——尤以束腰处的线刻『队戏图』与『巾舞图』和剔地起突的龙纹等耐人寻味，间以砂岩圆雕狮、虎、鹿主体的短柱，《营造法式》谓之『格身版柱』。再有，月台之上曾经建有献殿一座——证据则是后殿台帮正面的两方碑刻——金正隆三年（1158年）《举义□□仙□村重修献楼□□记》和大定三年（1163年）《举义□□□村砌基阶记》。感慨后代石匠的拮据和粗率之外，也庆幸雨水淋漓的岁月终究比不得不肖子孙的暴行。

学生们当时的工作只是开始，更多的疑问留给有心人未来继续探寻。初步介绍和回顾之后，作为编著者，我们还应当在此对测绘工作的核心成员表达衷心的感谢。他们分别是：崇明寺测绘的带队老师贺从容、游仙寺的带队老师贾珺、开化寺的带队老师李路珂和西李门二仙庙的带队老师王贵祥。此外，清华大学建筑学院建筑历史与文物保护研究所的博士、硕士研究生也一直在核心团队中扮演不可或缺的角色。不同的年份他们会出现在不同的遗产地的工作现场，不同的岗位，担当学习、研究和辅导员的工作。请允许我们在这里记录他们的姓名：（韩）辛惠园、（韩）黄文镐、（韩）朴沼衍、杨澍、张亦驰、翁帆、李沁园、孙蕾、徐腾、赵萨日娜、傅娜、王曦晨、何文轩、刘梦雨、姜铮、赵寿堂、徐扬、蒋哲、刘潇潇、赵波。

刘畅
清华大学建筑学院 建筑历史与文物保护研究所

图版

Figure

崇明寺
Chongming Monastery

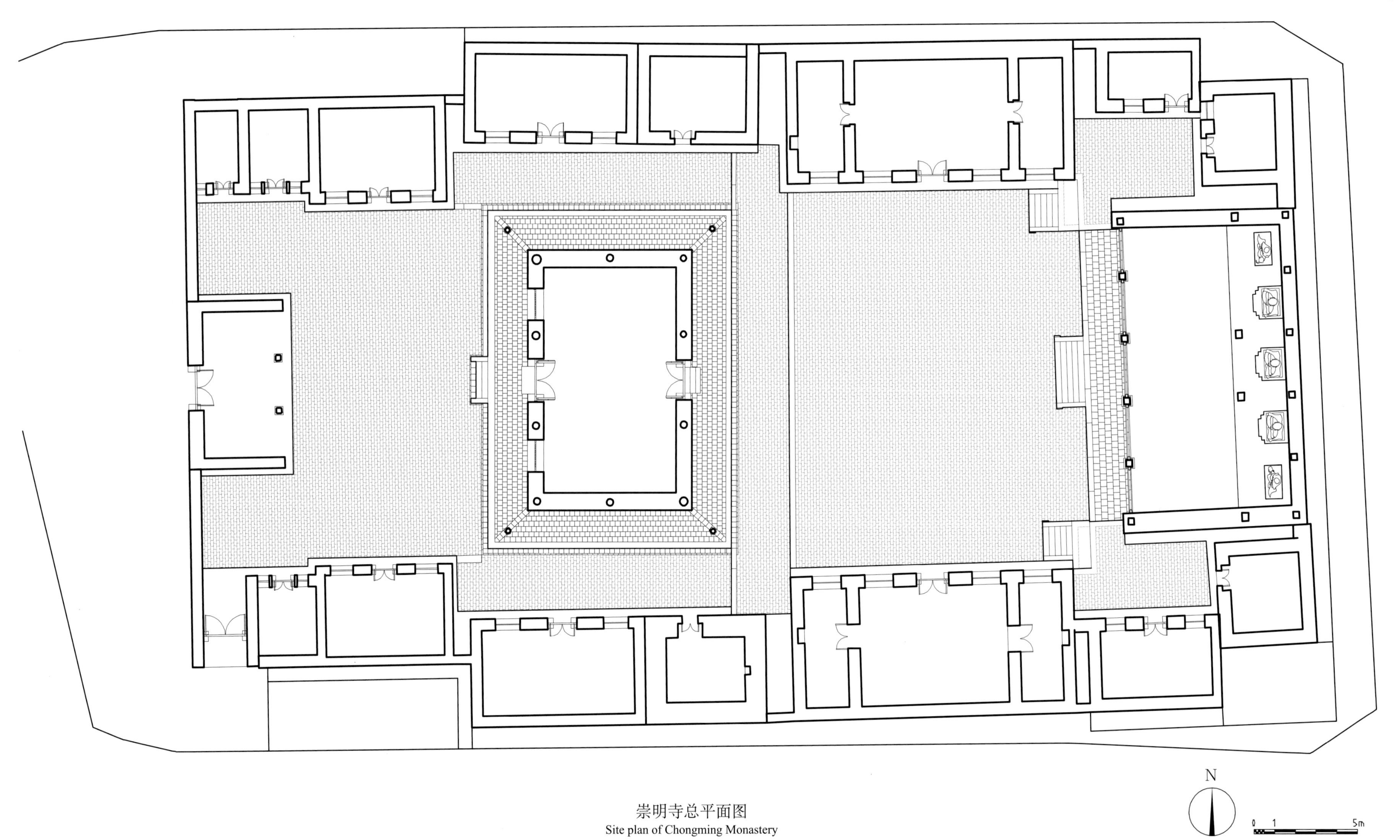

崇明寺总平面图
Site plan of Chongming Monastery

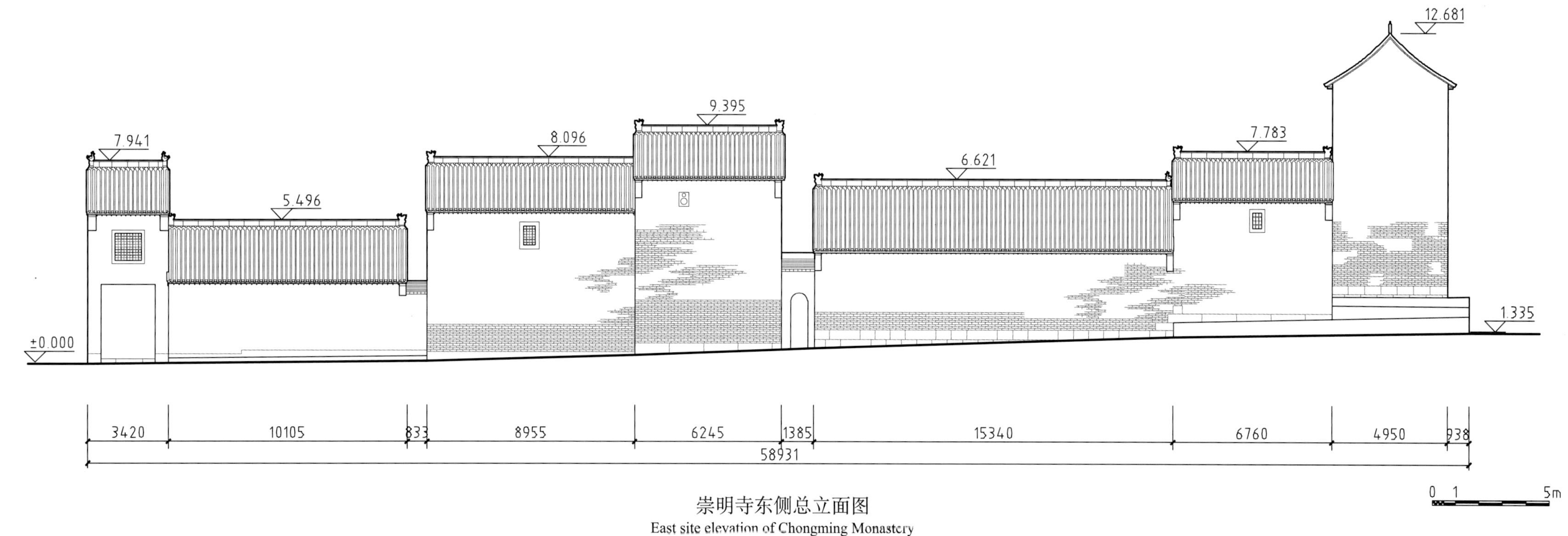

崇明寺东侧总立面图
East site elevation of Chongming Monastery

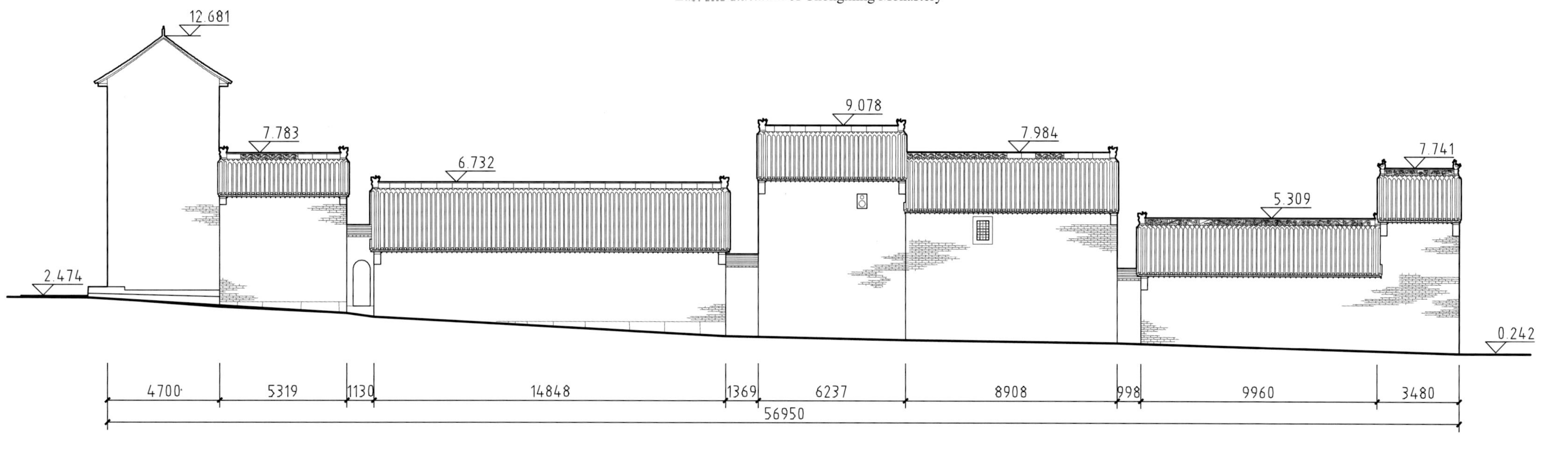

崇明寺西侧总立面图
West site elevation of Chongming Monastery

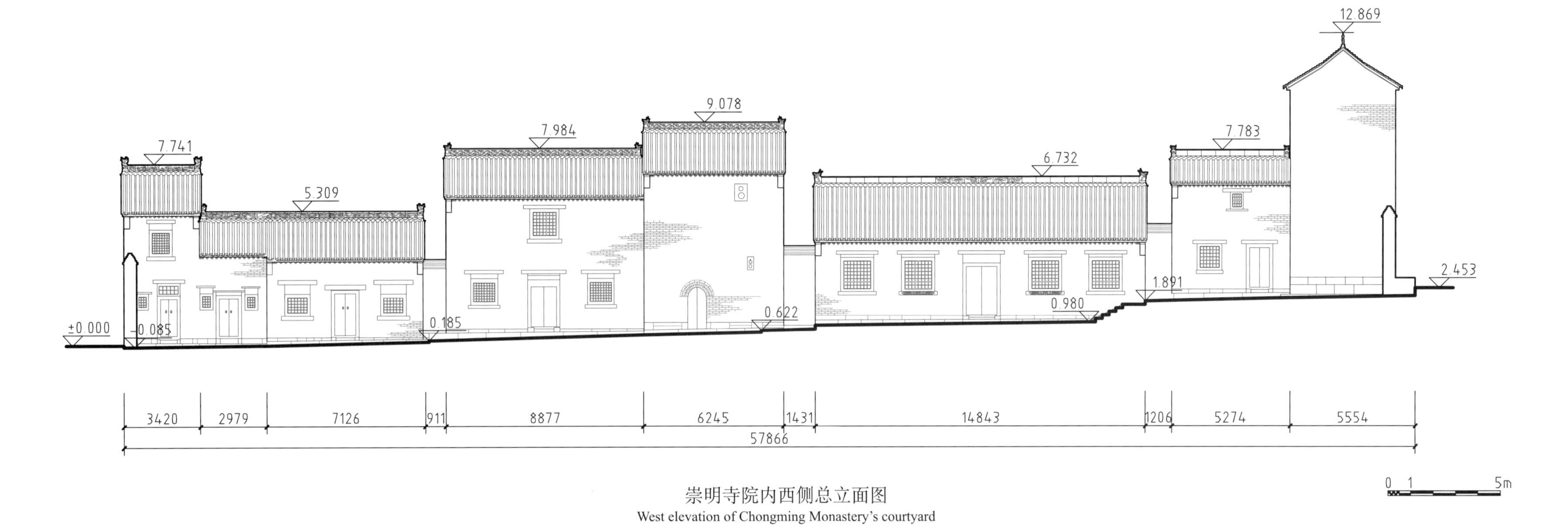

崇明寺院内西侧总立面图

West elevation of Chongming Monastery's courtyard

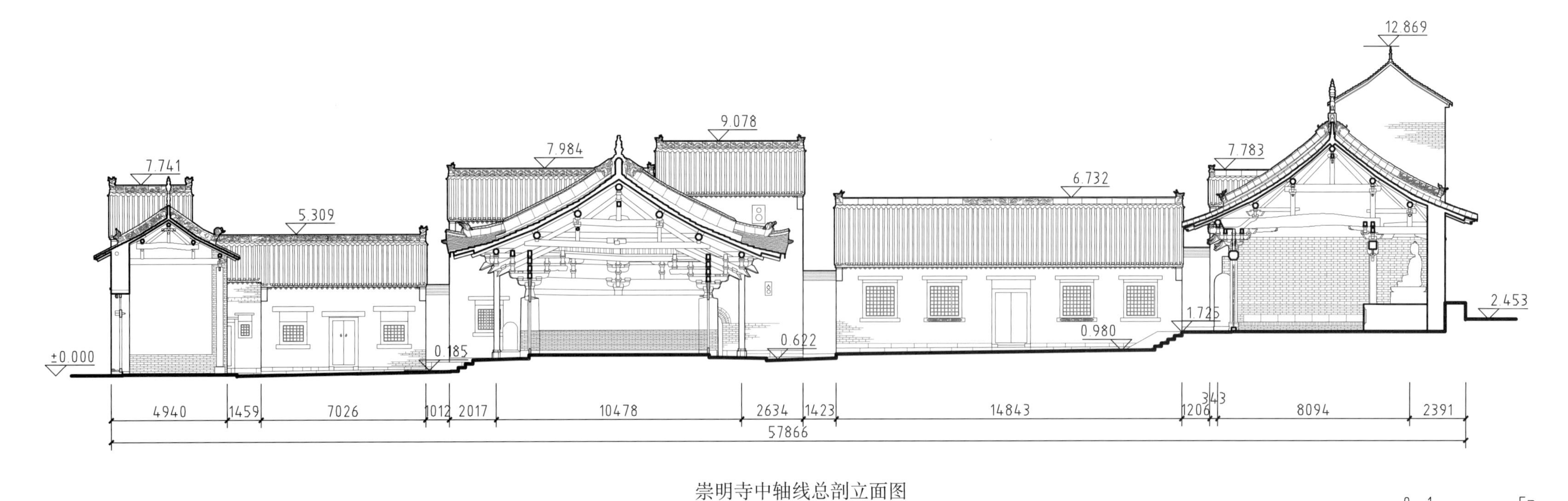

崇明寺中轴线总剖立面图

Site sectional elevation of Chongming Monastery's central axis

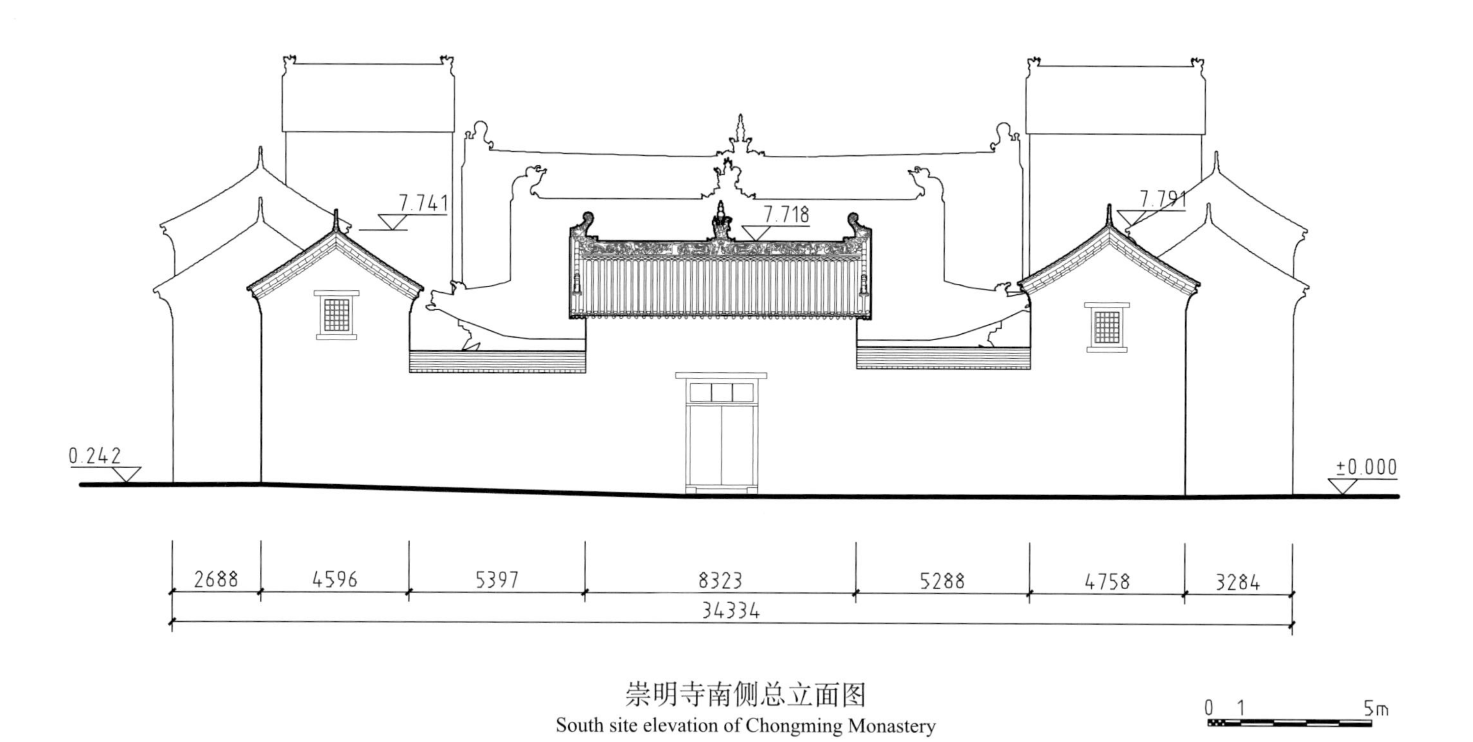

崇明寺南侧总立面图

South site elevation of Chongming Monastery

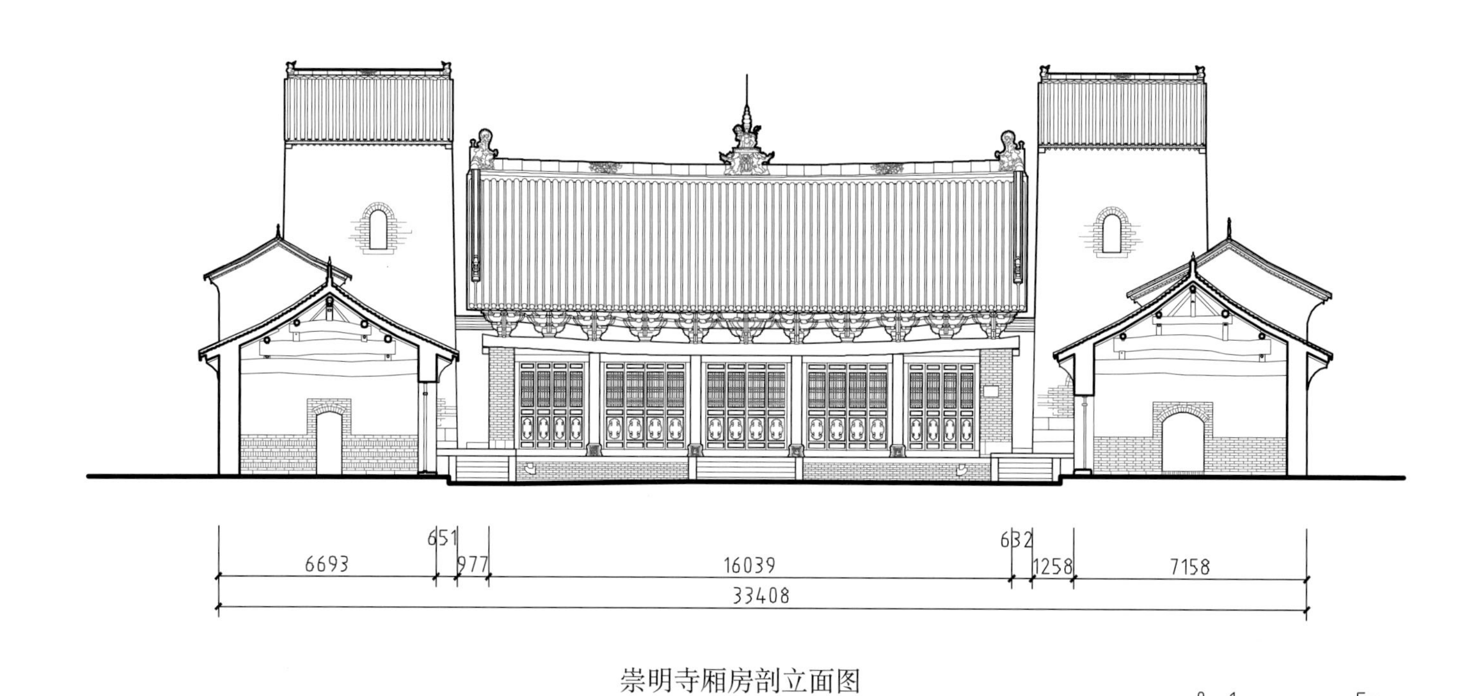

崇明寺厢房剖立面图

Sectional elevation of Chongming Monastery's wing-room

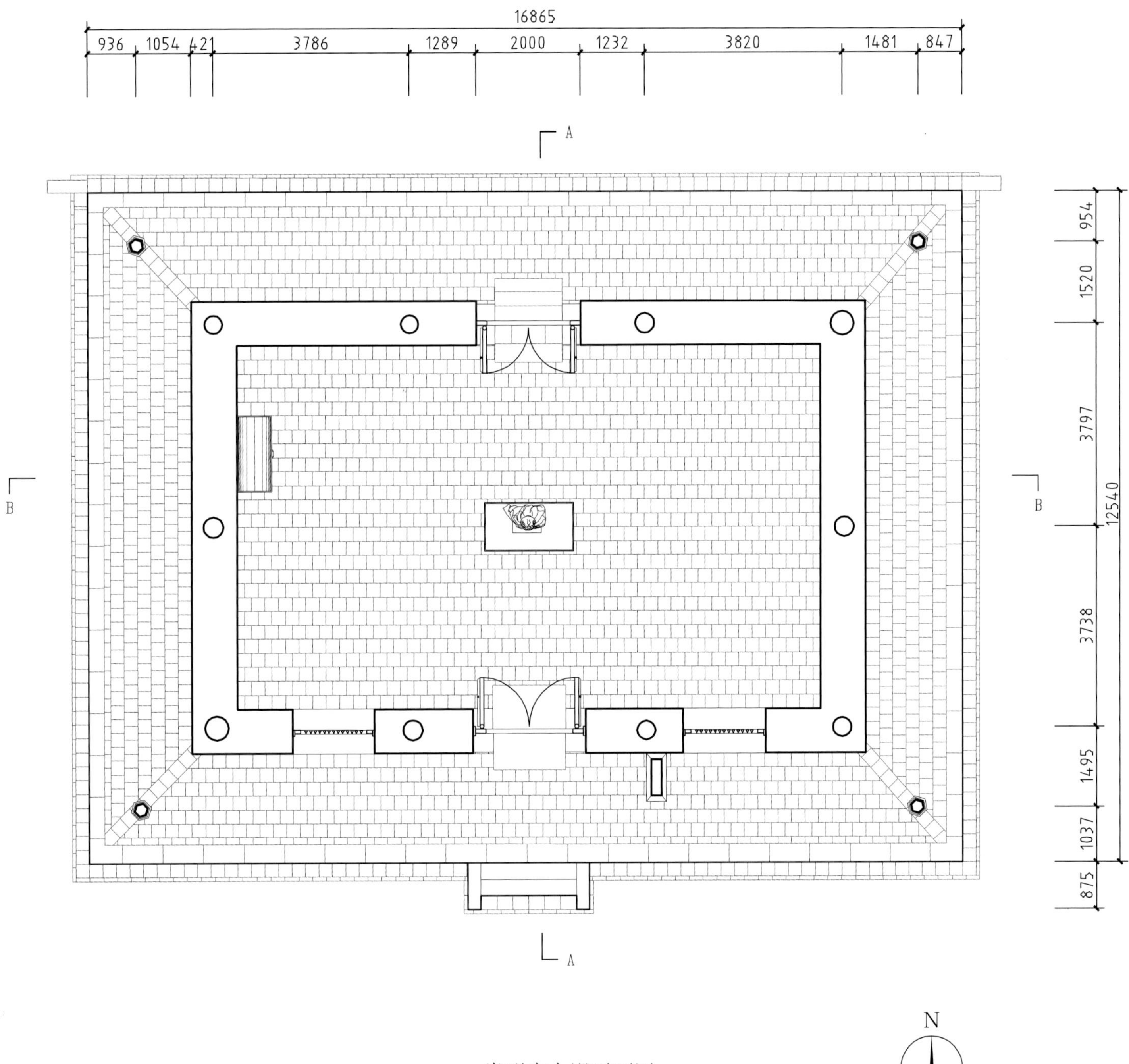

崇明寺大殿平面图
Plan of Chongming Monastery's main hall

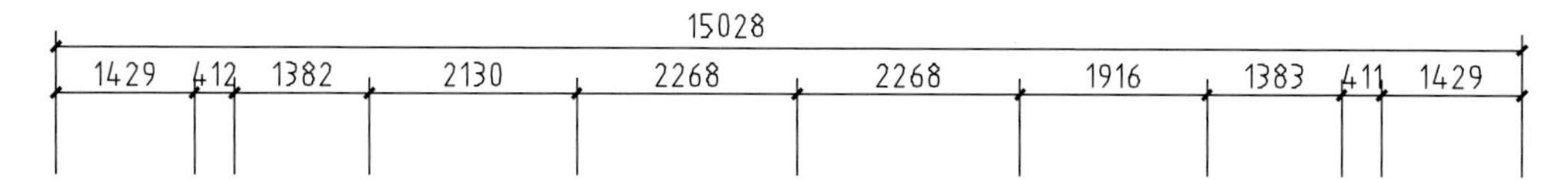

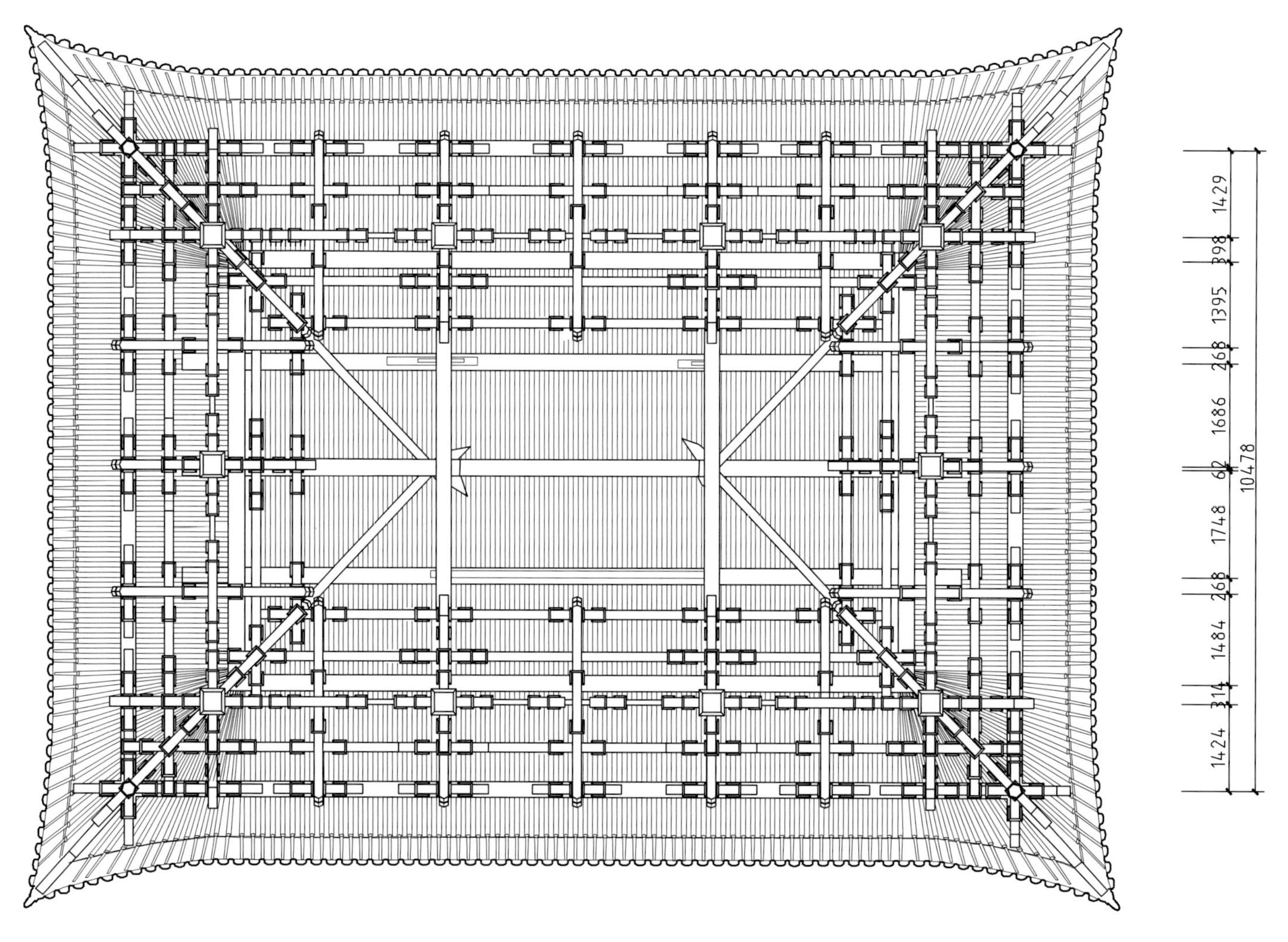

崇明寺大殿屋架仰视图

Plan of Chongming Monastery's main hall's roof framework as seen from below

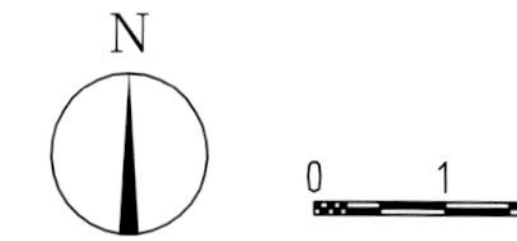

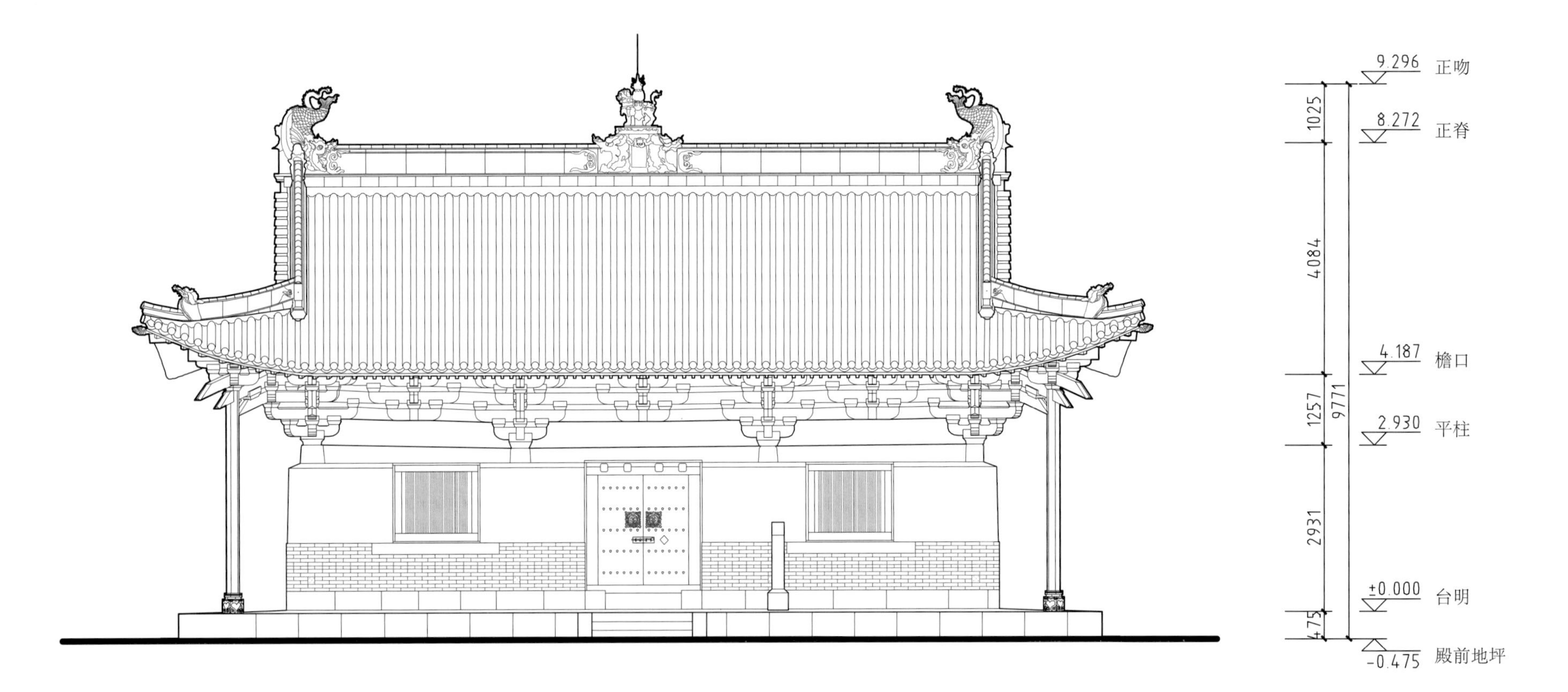

崇明寺大殿正立面图

Front elevation of Chongming Monastery's main hall

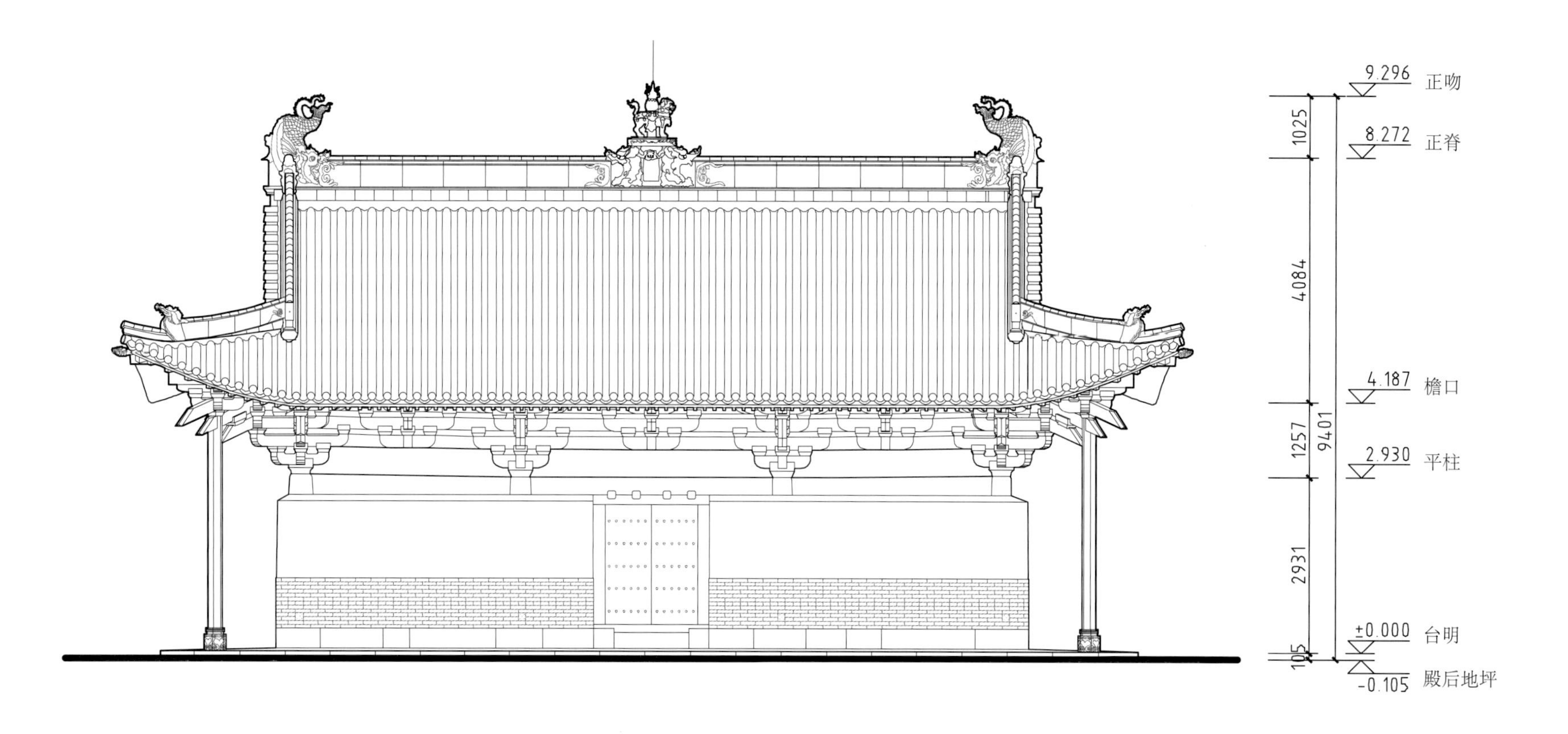

崇明寺大殿背立面图

Rear elevation of Chongming Monastery's main hall

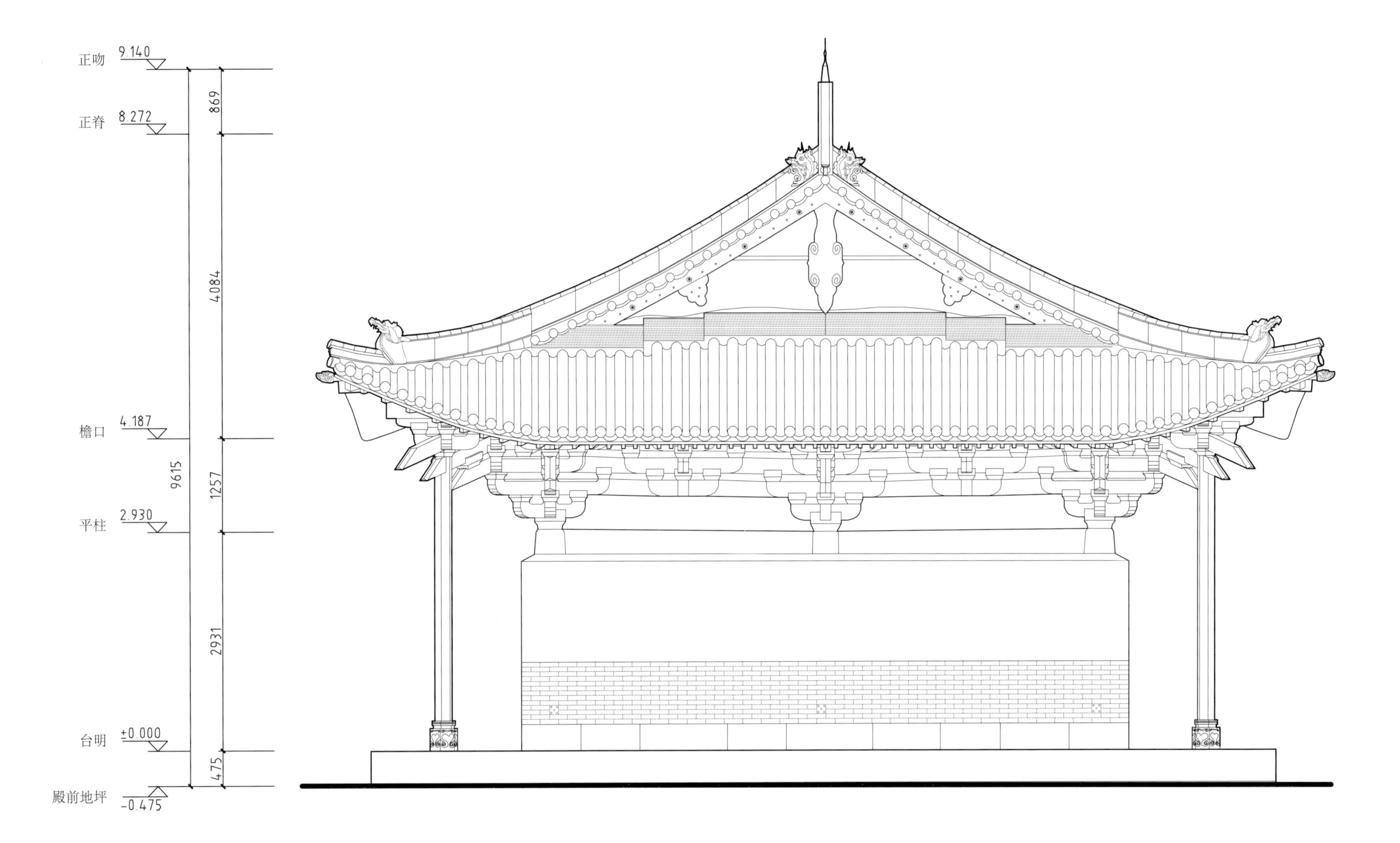

崇明寺大殿侧立面图

Side elevation of Chongming Monastery's main hall

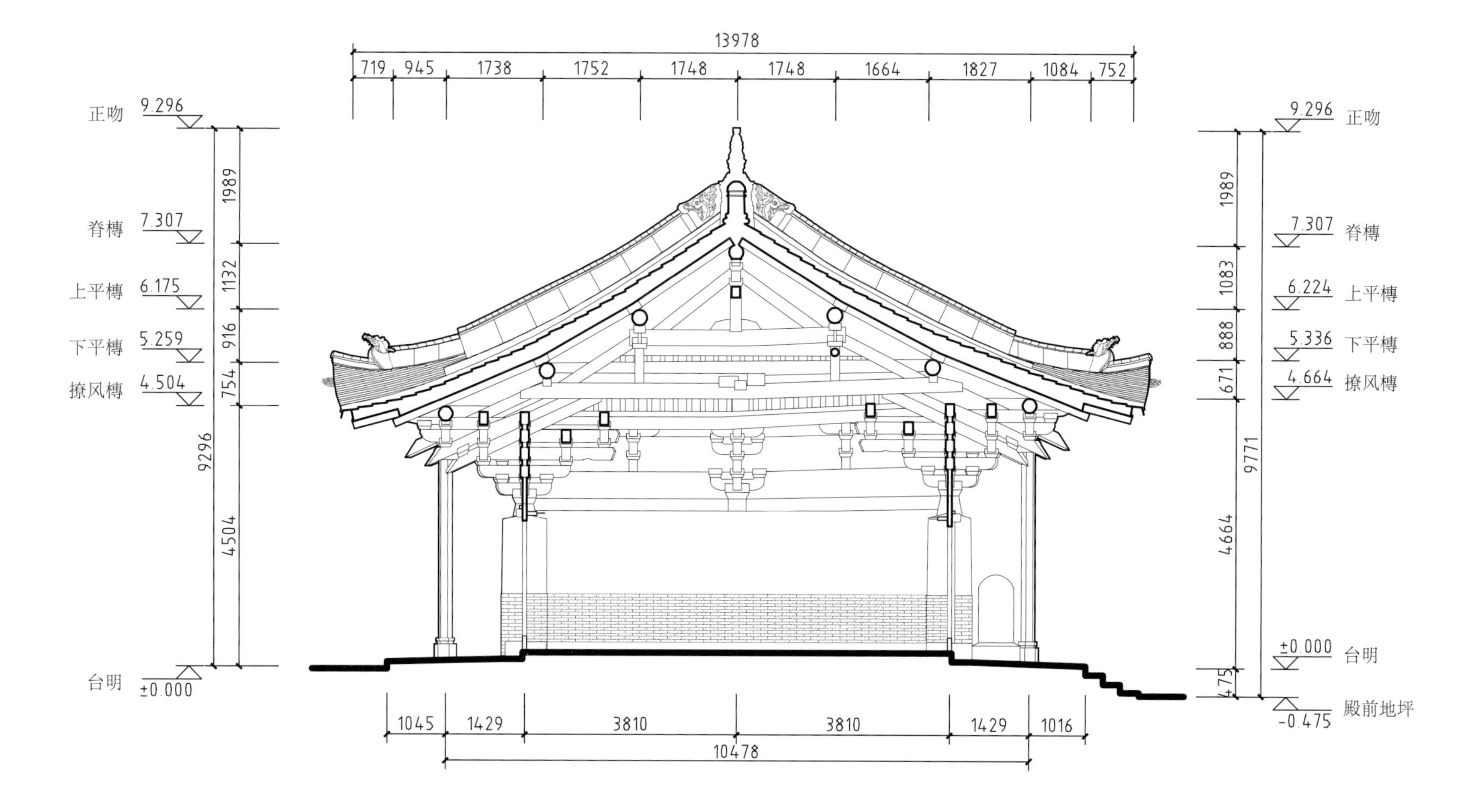

崇明寺大殿 A-A 剖面图

Section A-A of Chongming Monastery's main hall

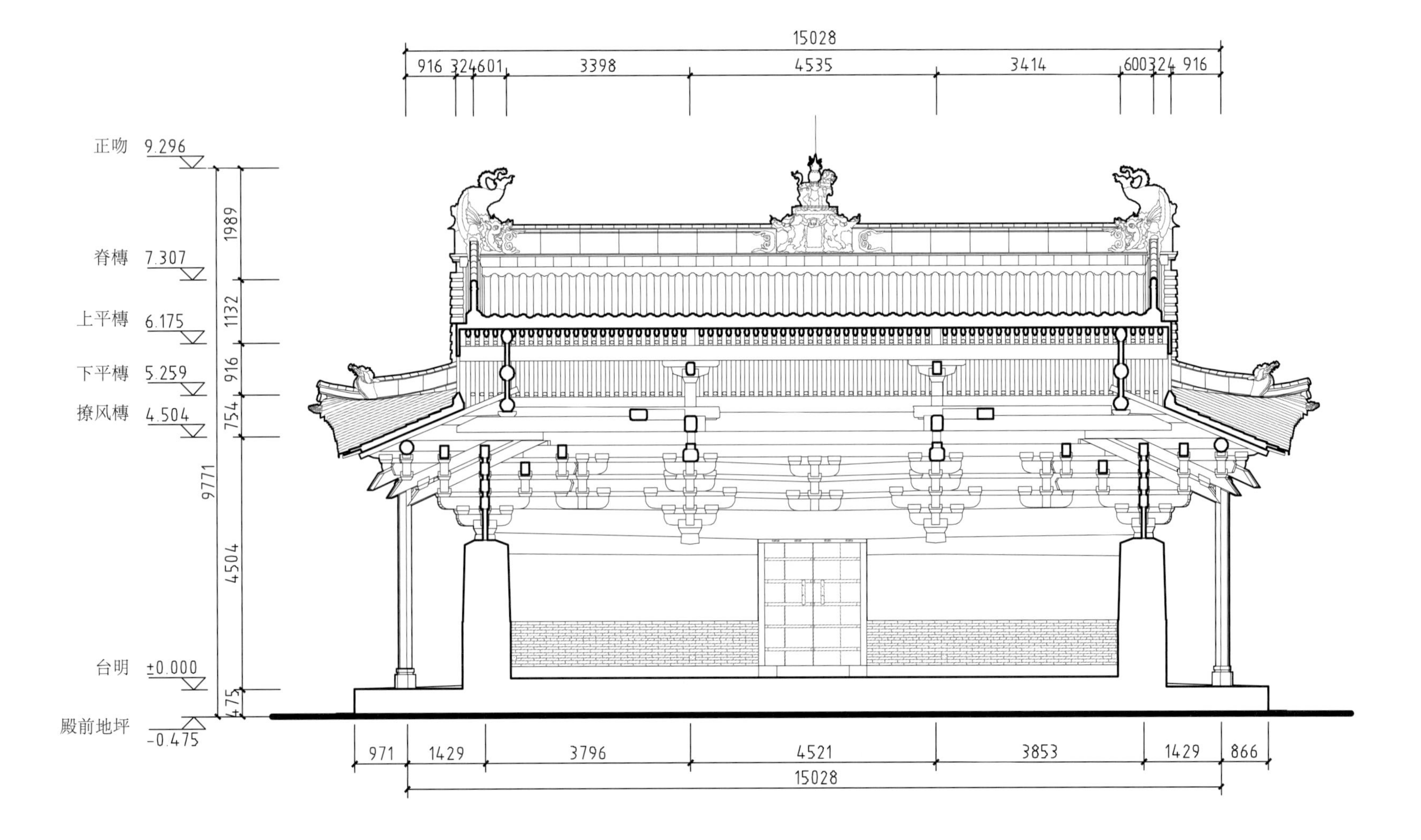

崇明寺大殿 B-B 剖面图
Section B-B of Chongming Monastery's main hall

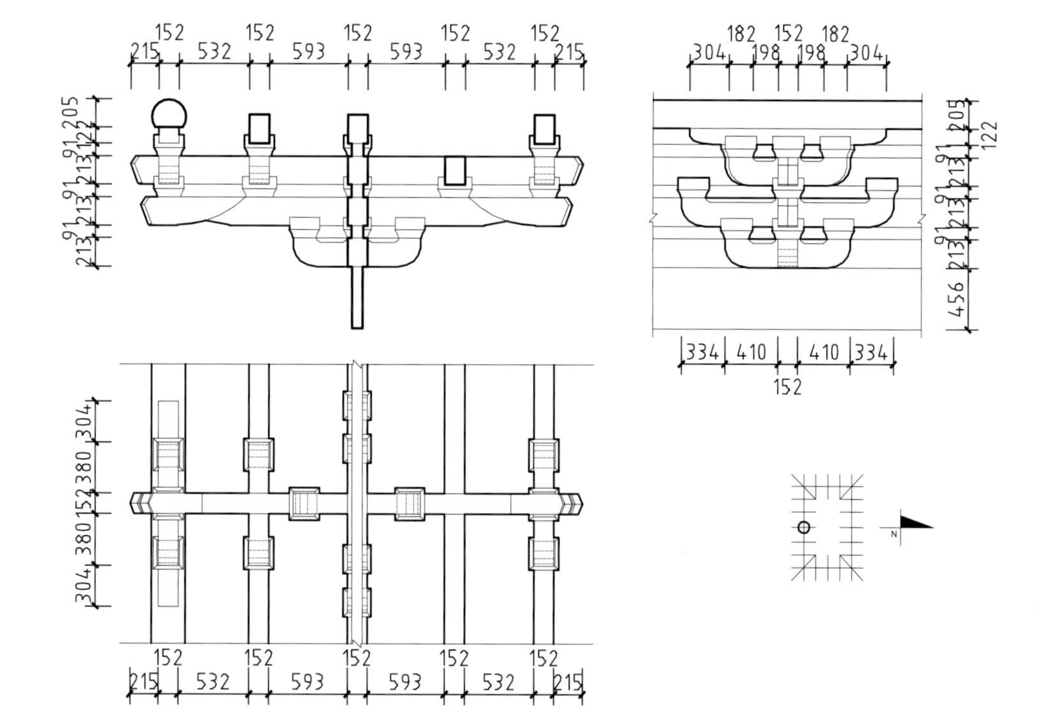

崇明寺大殿补间斗栱大样图
Intercolumnar bracket set of Chongming Monastery's main hall

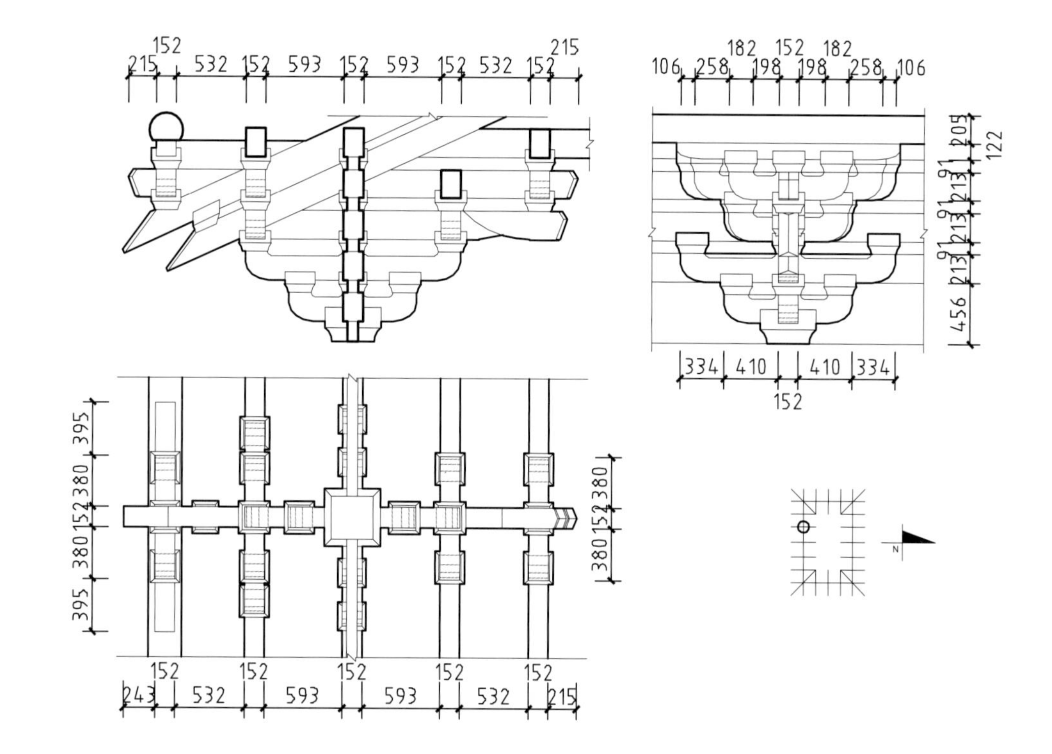

崇明寺大殿柱头斗栱大样图
Column-top bracket set of Chongming Monastery's main hall

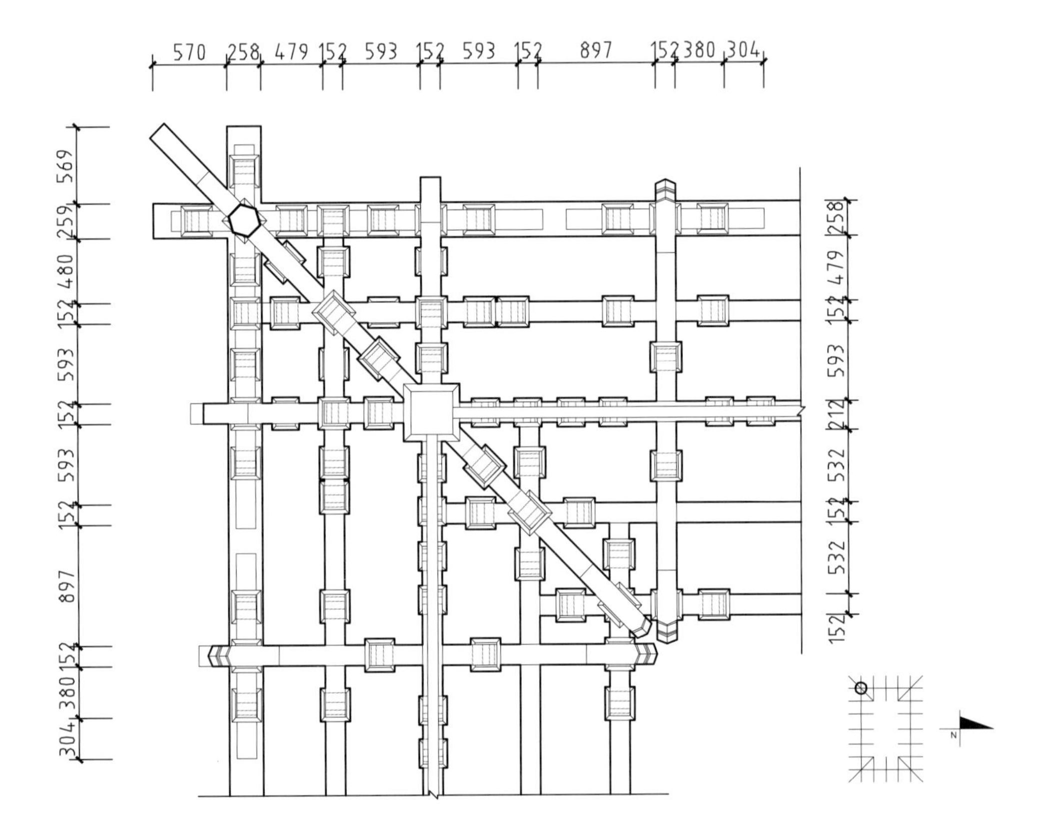

崇明寺大殿转角斗栱仰视图

Corner bracket set of Chongming Monastery's main hall as seen from below

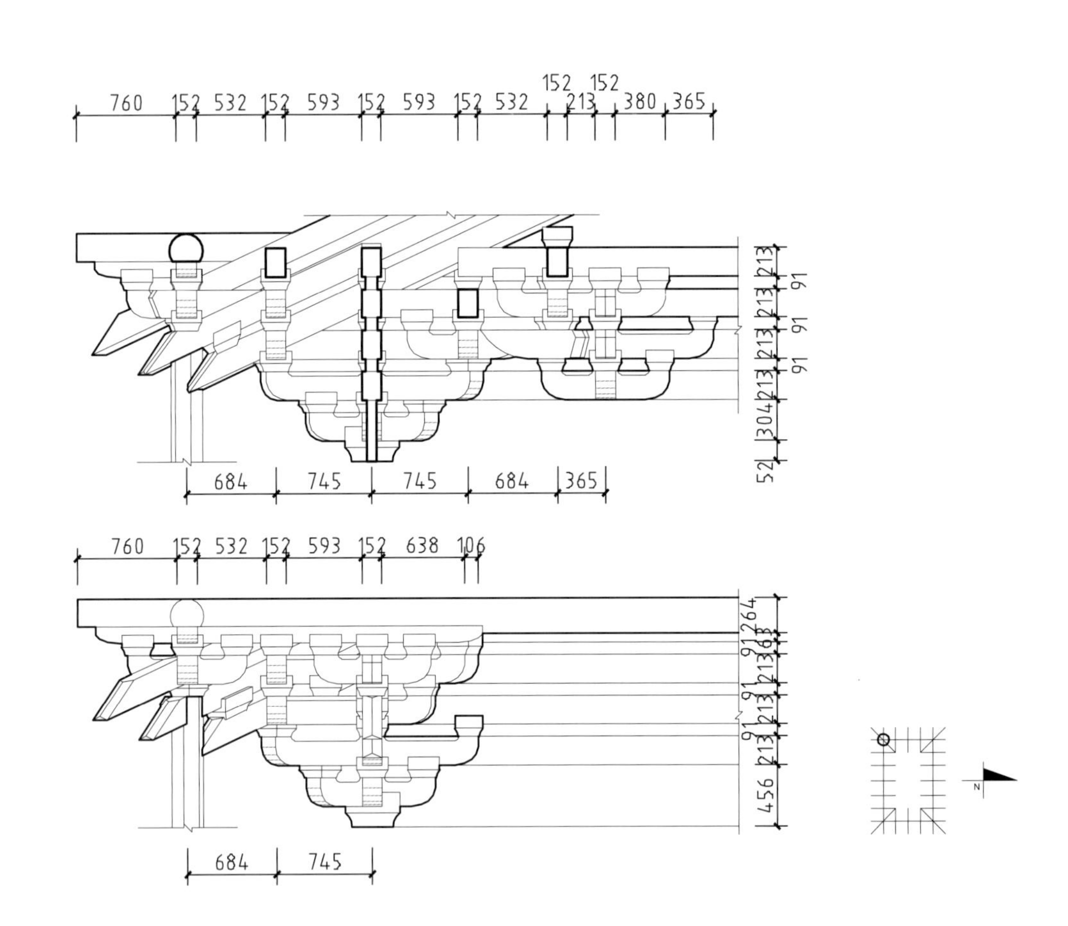

崇明寺大殿柱头斗栱剖面、立面图

Section and elevation of Chongming Monastery's main hall's column-top bracket set

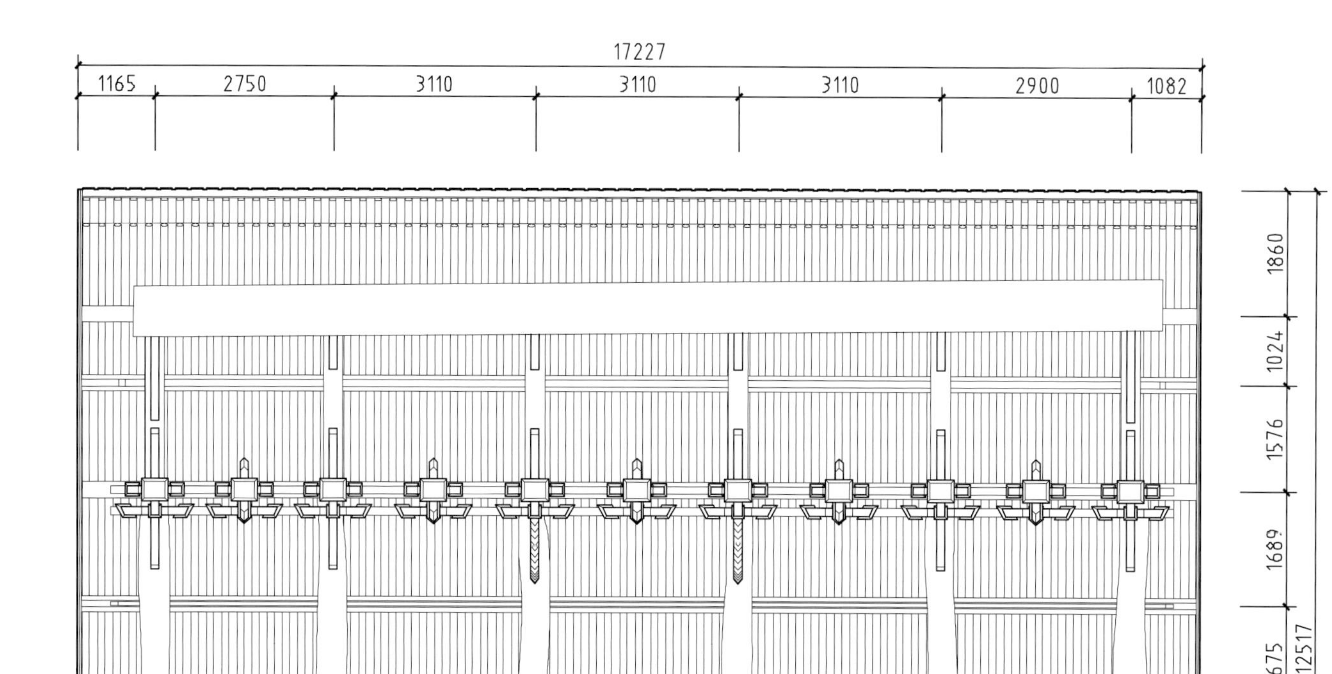

崇明寺后殿梁架仰视平面图

Plan of Chongming Monastery's rear hall's framework as seen from below

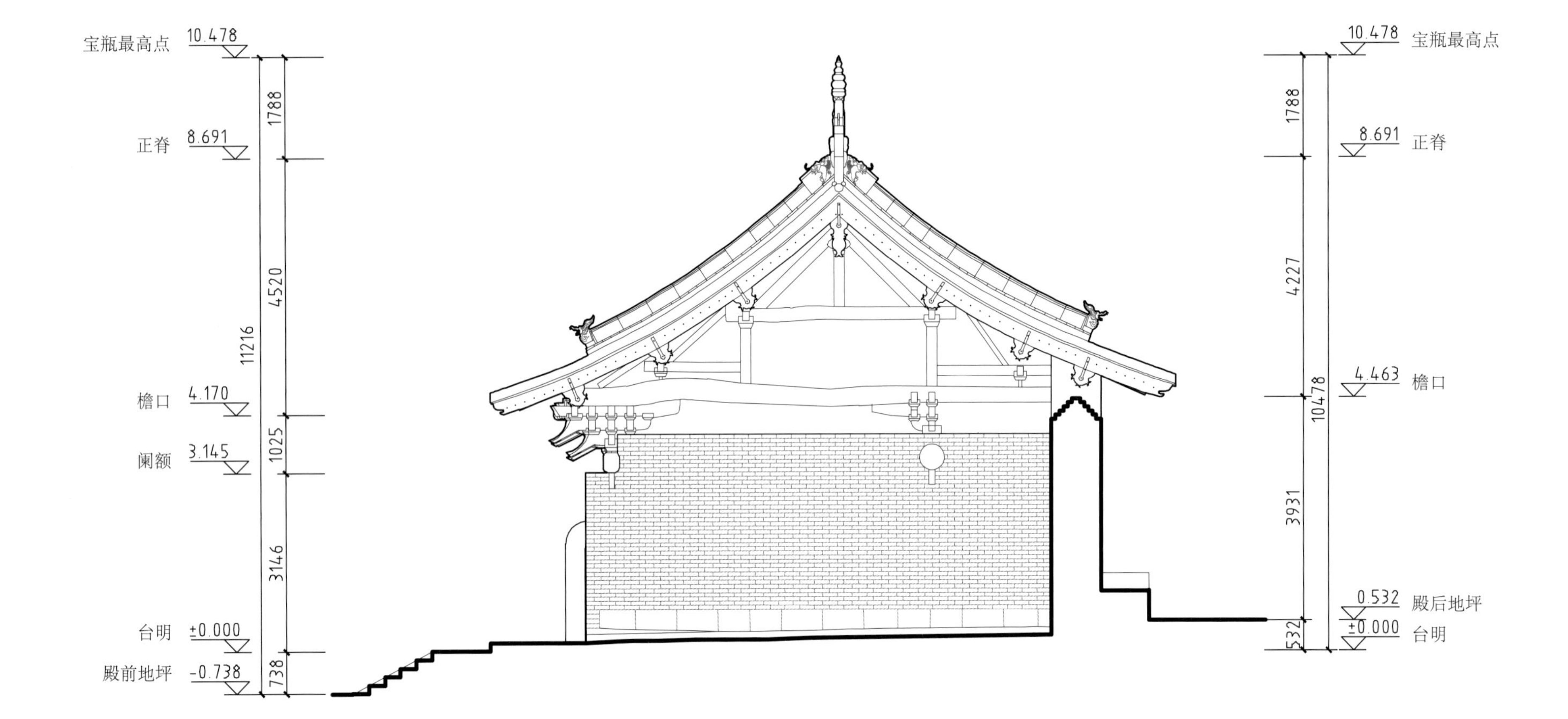

崇明寺后殿侧立面图
Side elevation of Chongming Monastery's rear hall

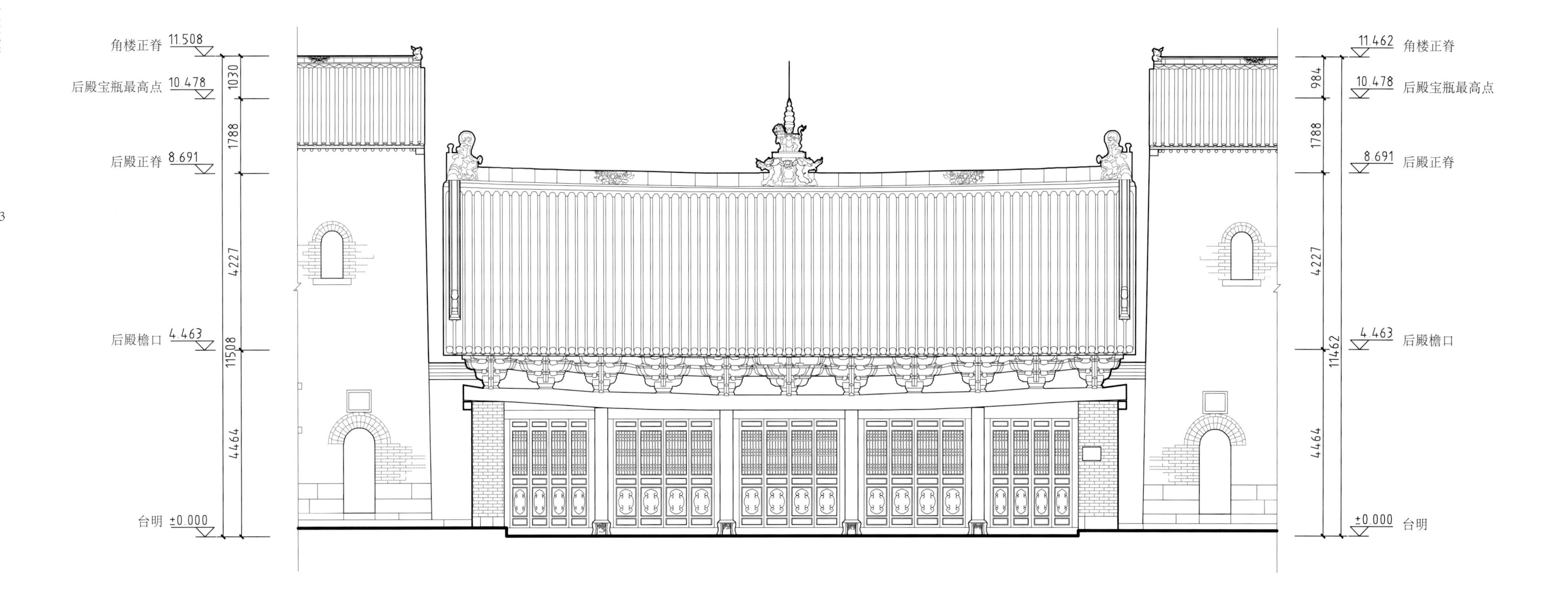

崇明寺后殿正立面图

Front elevation of Chongming Monastery's rear hall

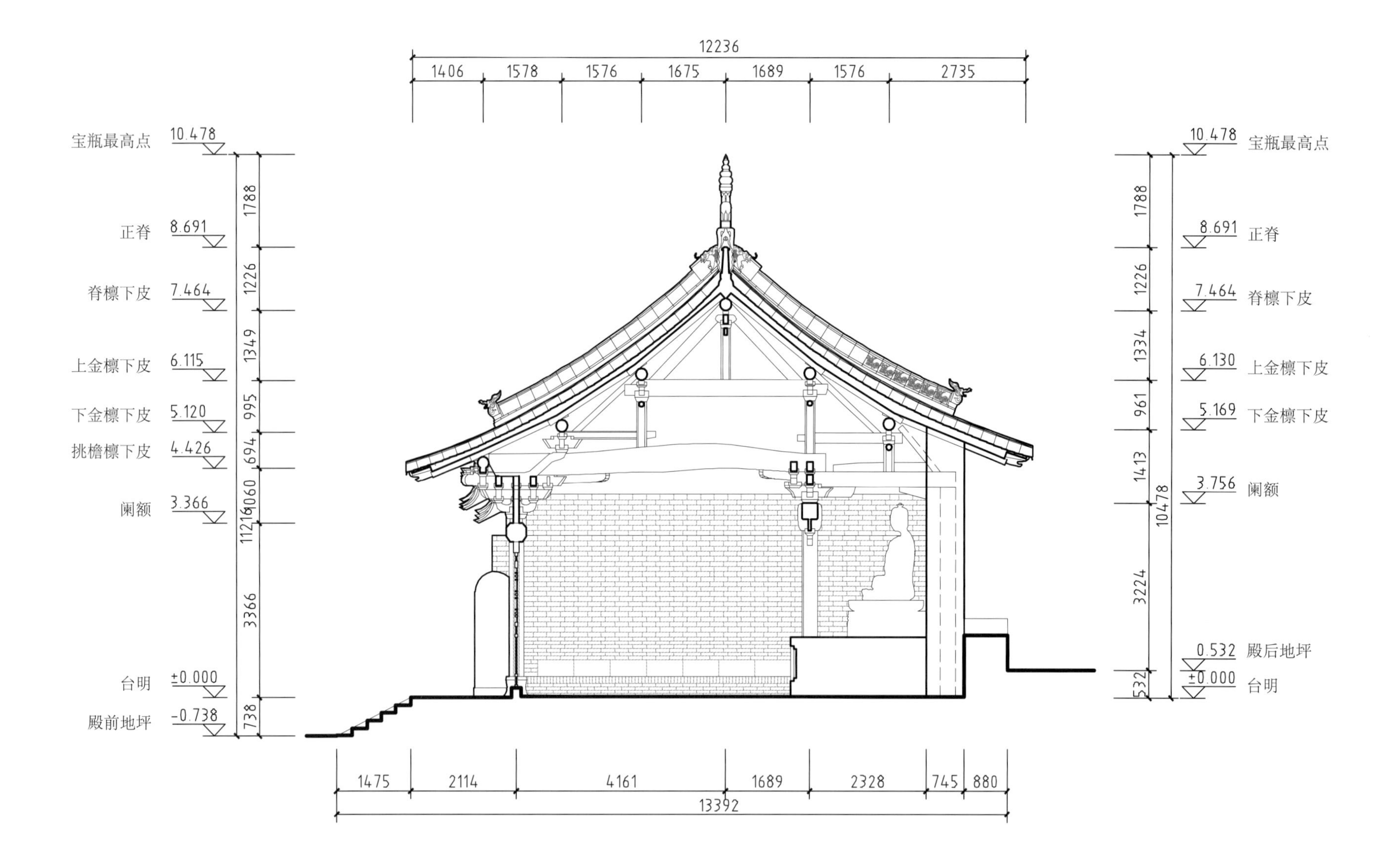

崇明寺后殿明间横剖面图

Cross-section of Chongming Monastery's rear hall's central-bay

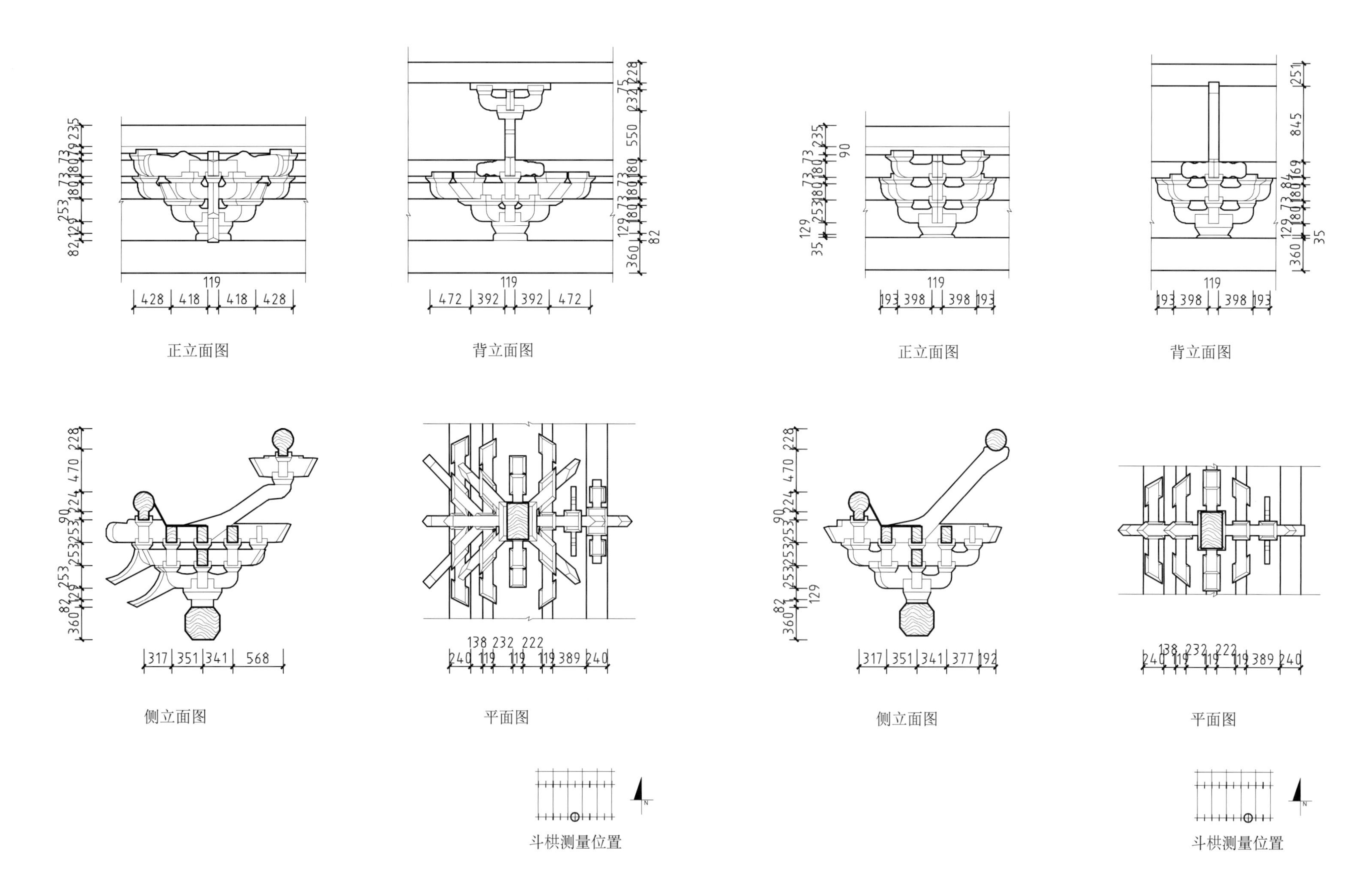

崇明寺后殿明间平身科斗栱大样图
Intercolumnar bracket set of Chongming Monastery's rear hall's central-bay

崇明寺后殿次间平身科斗栱大样图
Intercolumnar bracket set of Chongming Monastery's rear hall's side-bay

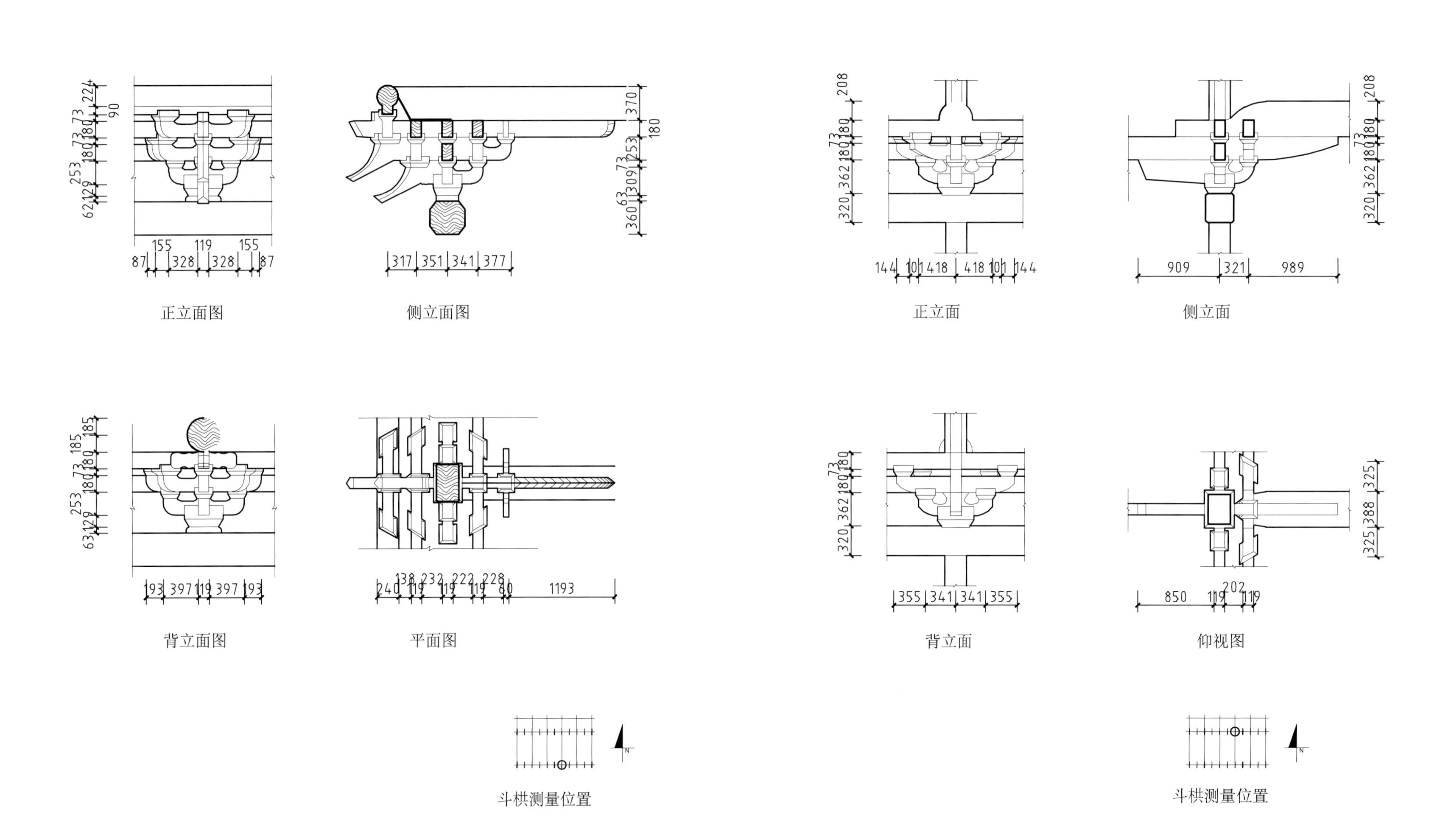

崇明寺后殿明间柱头科斗栱大样图
Column-top bracket set of Chongming Monastery's rear hall's central-bay

崇明寺后殿殿内柱头科斗栱大样图
Column-top bracket set inside Chongming Monastery's rear hall

游仙寺
Youxian Monastery

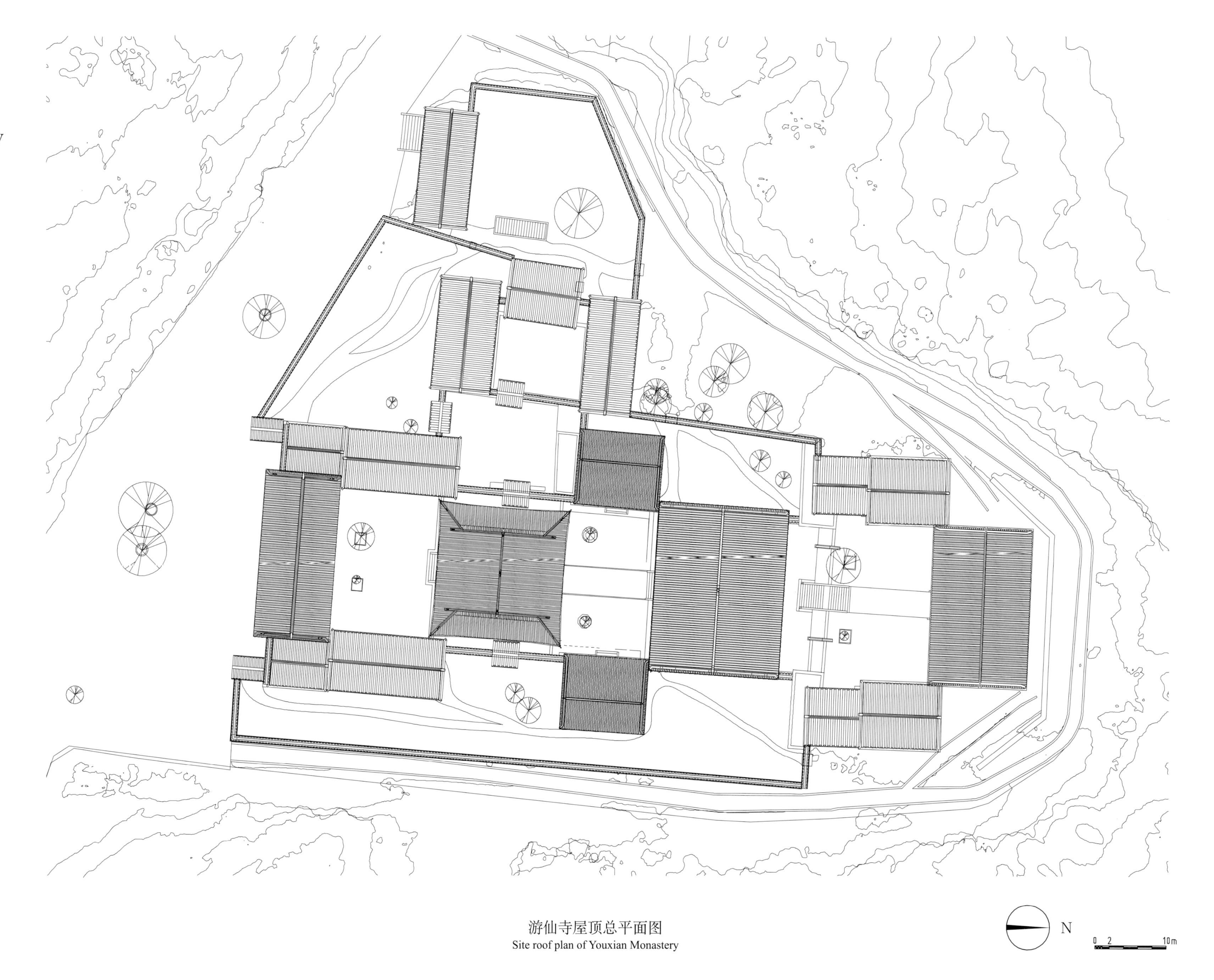

游仙寺屋顶总平面图
Site roof plan of Youxian Monastery

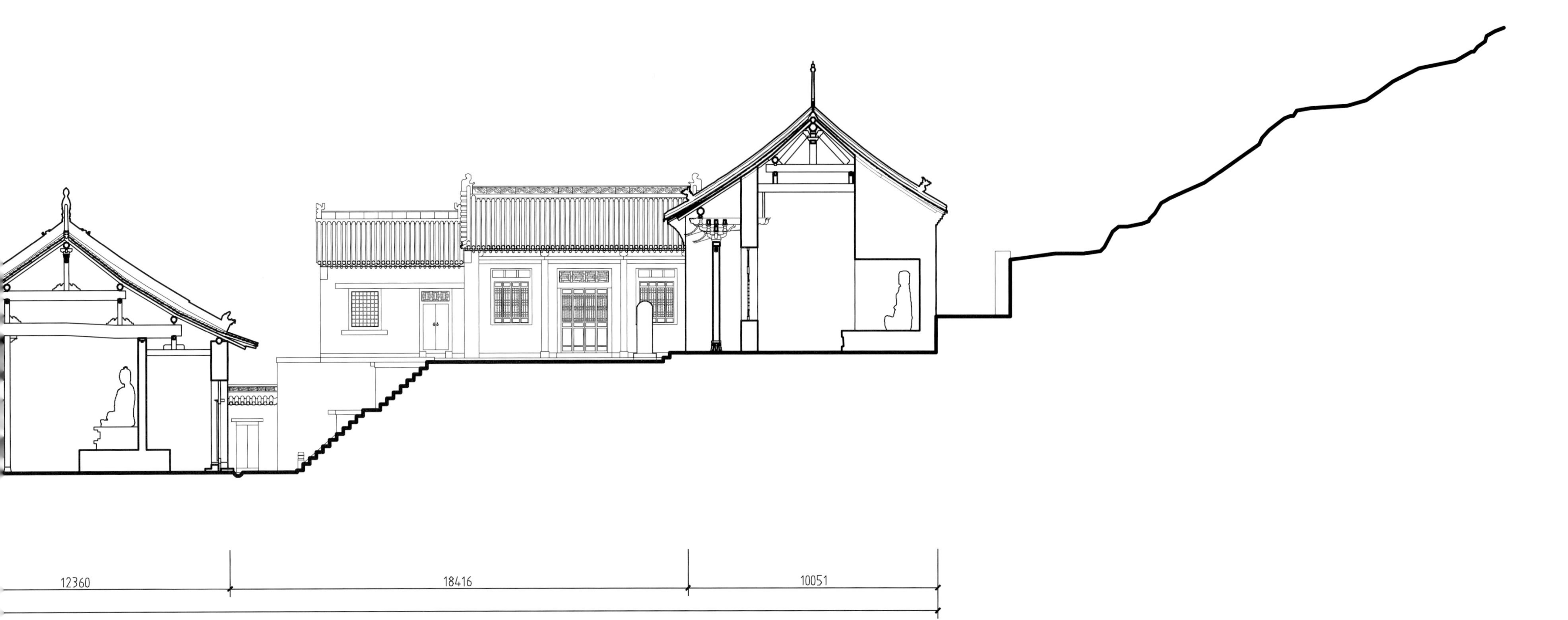
12360
18416
10051

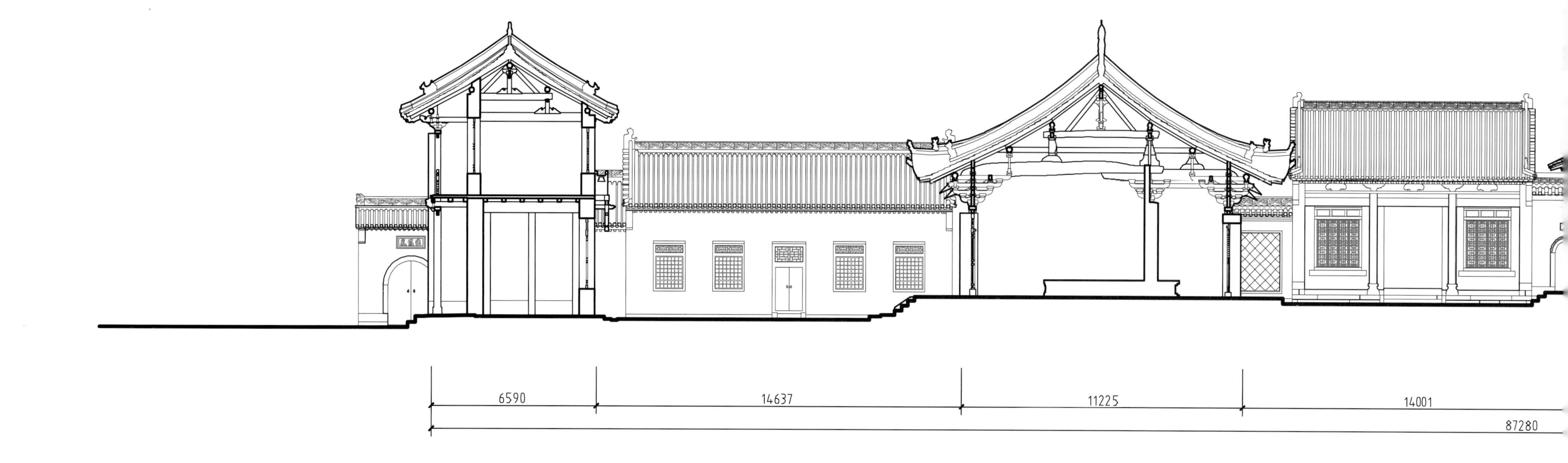

游仙寺总纵剖面图

Longitudinal site section of Youxian Monastery

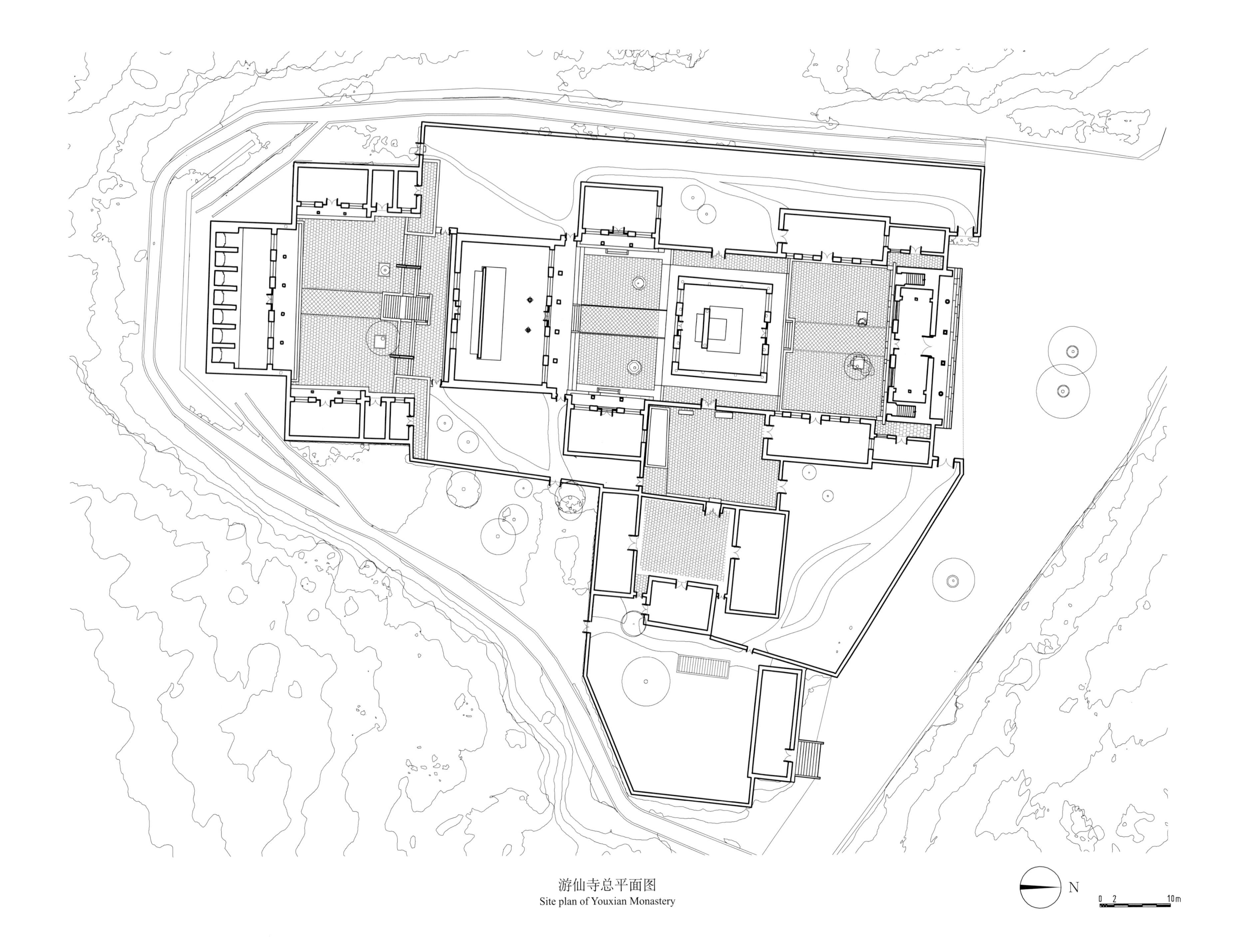

游仙寺总平面图
Site plan of Youxian Monastery

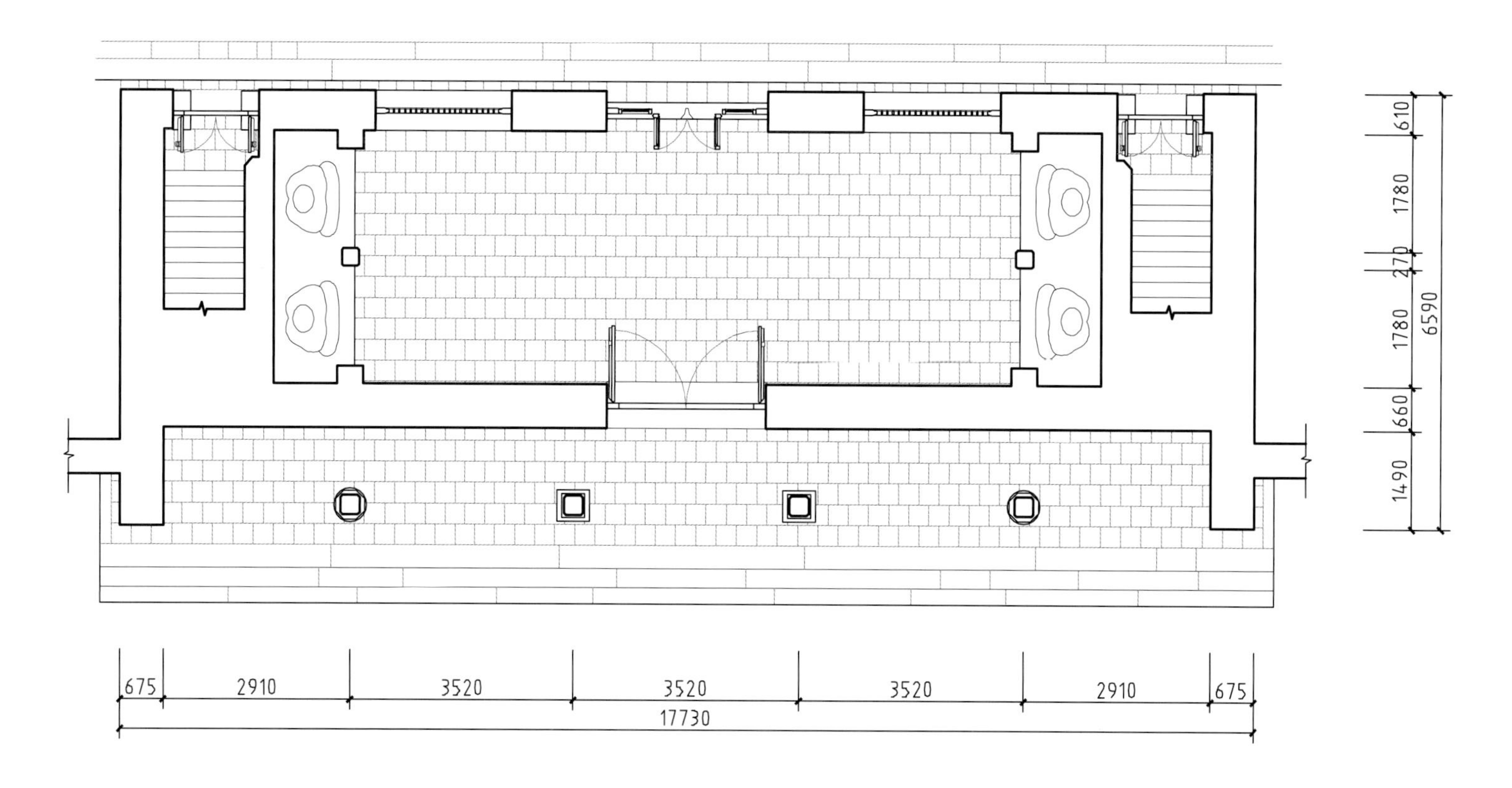

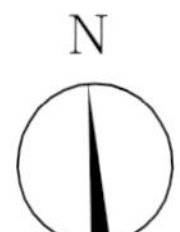

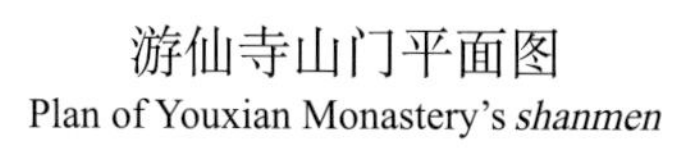
游仙寺山门平面图
Plan of Youxian Monastery's *shanmen*

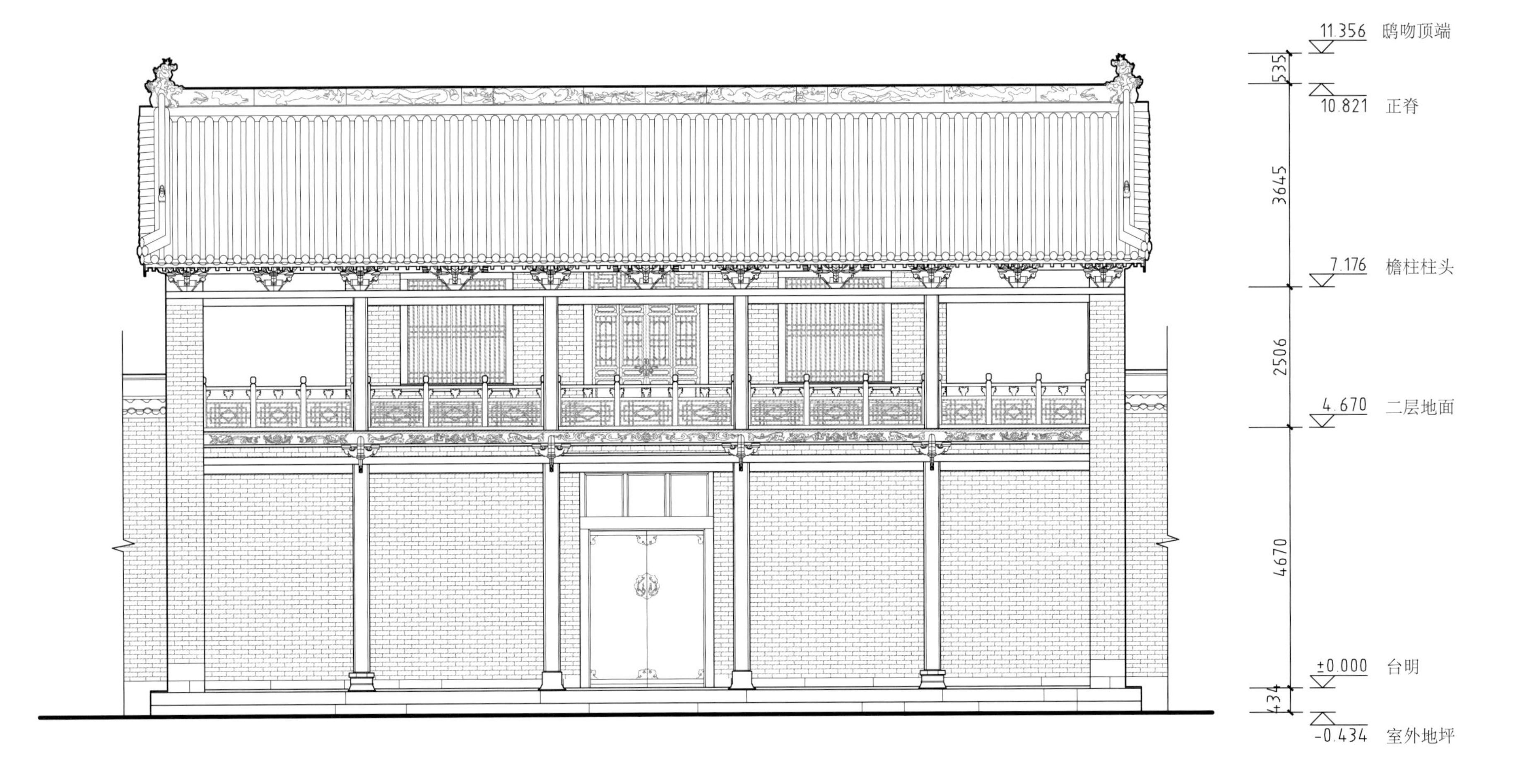

游仙寺山门正立面图

Front elevation of Youxian Monastery's *shanmen*

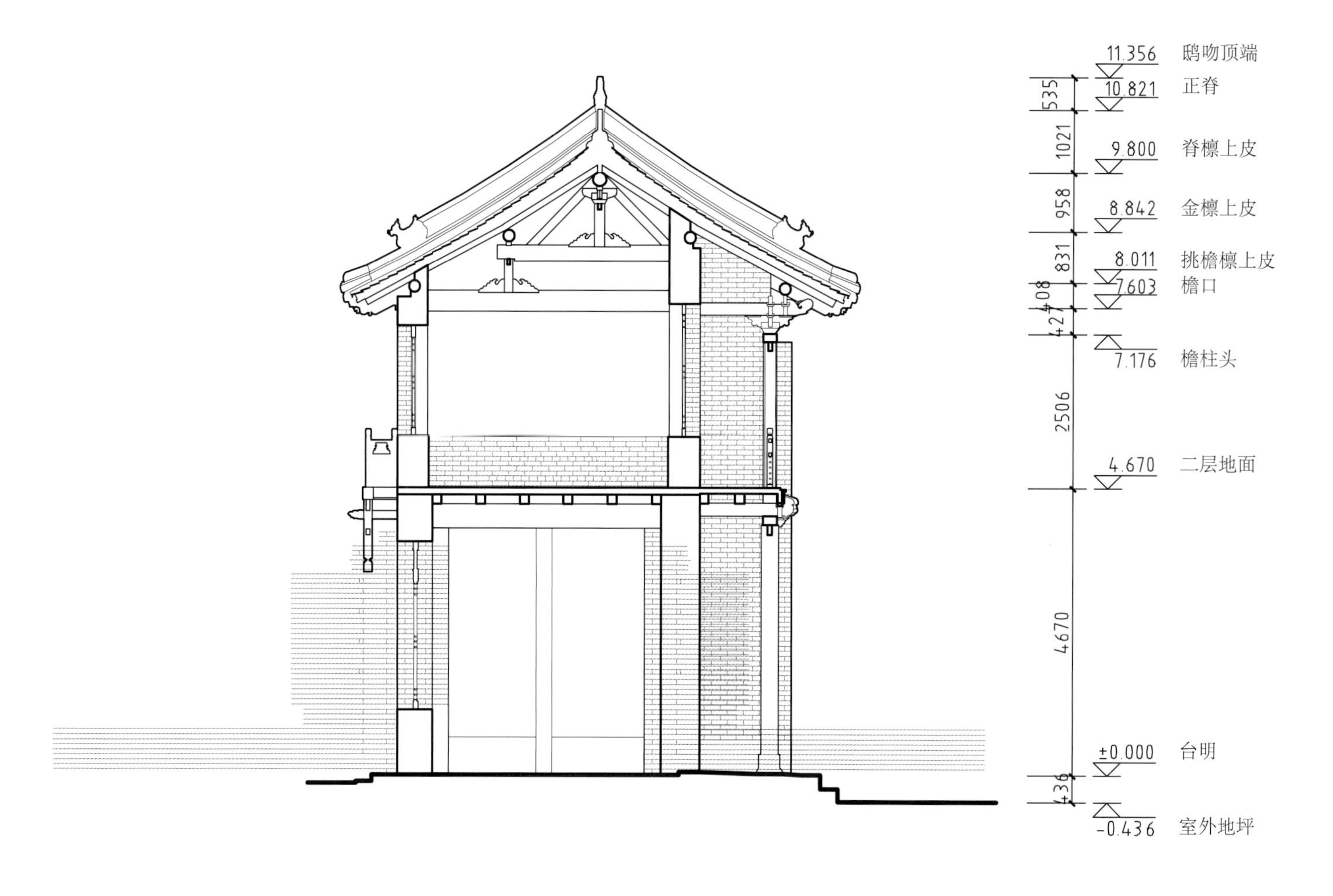

游仙寺山门横剖面图

Cross-section of Youxian Monastery's *shanmen*

游仙寺山门纵剖面图

Longitudinal section of Youxian Monastery's *shanmen*

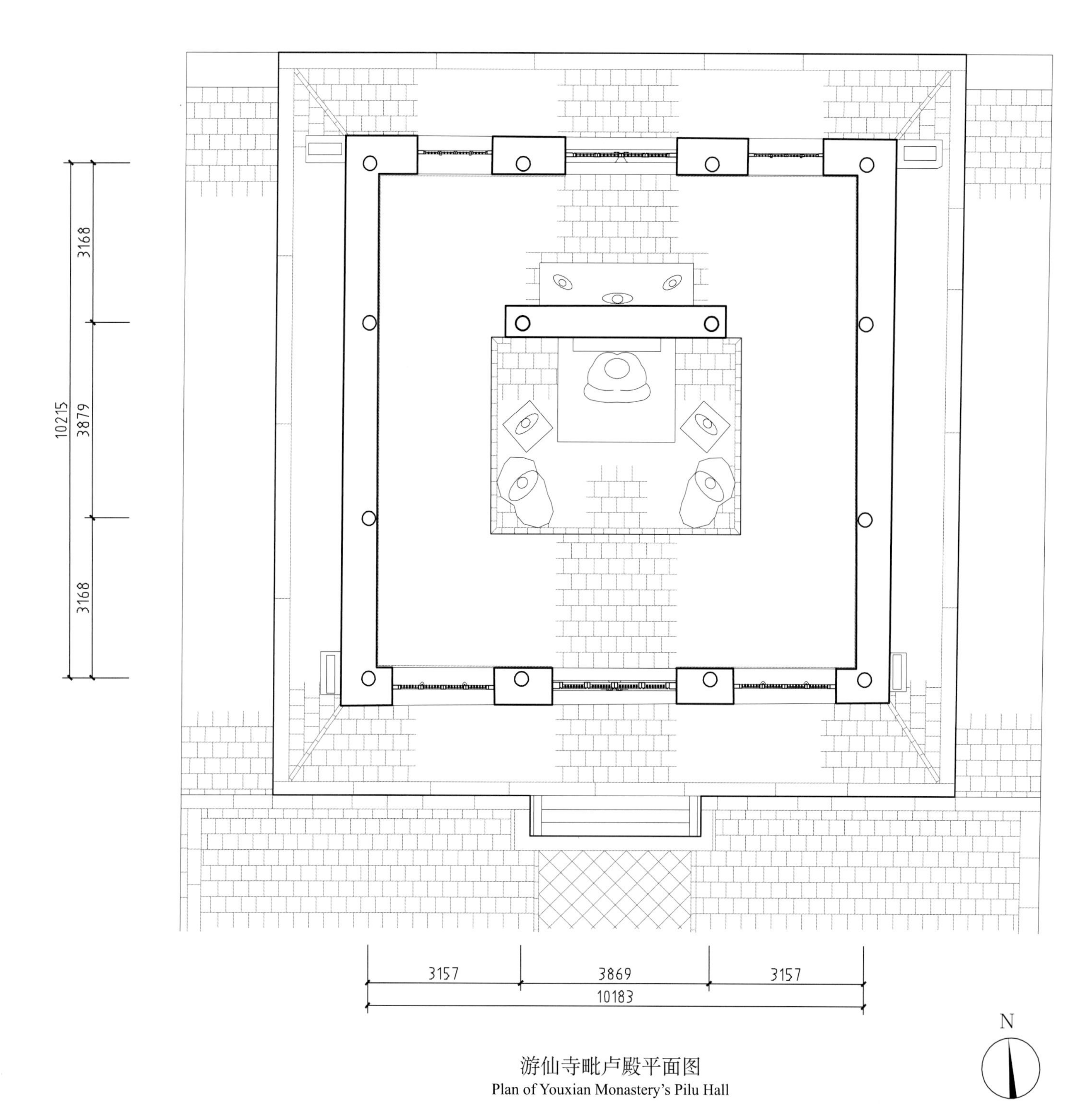

游仙寺毗卢殿平面图
Plan of Youxian Monastery's Pilu Hall

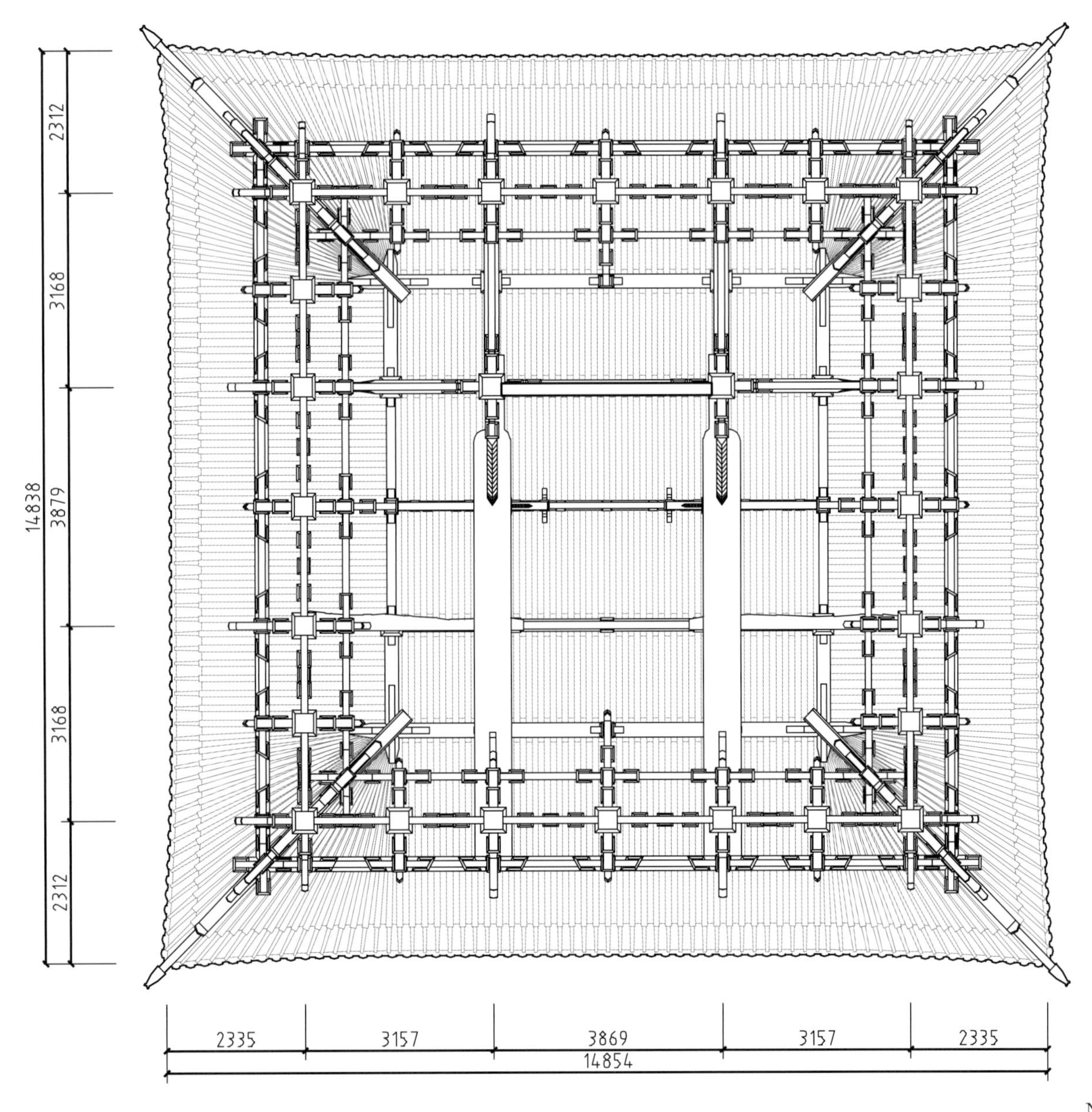

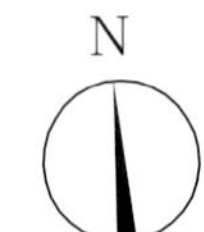

游仙寺毗卢殿梁架仰视平面图

Plan of Youxian Monastery's Pilu Hall's framework as seen from below

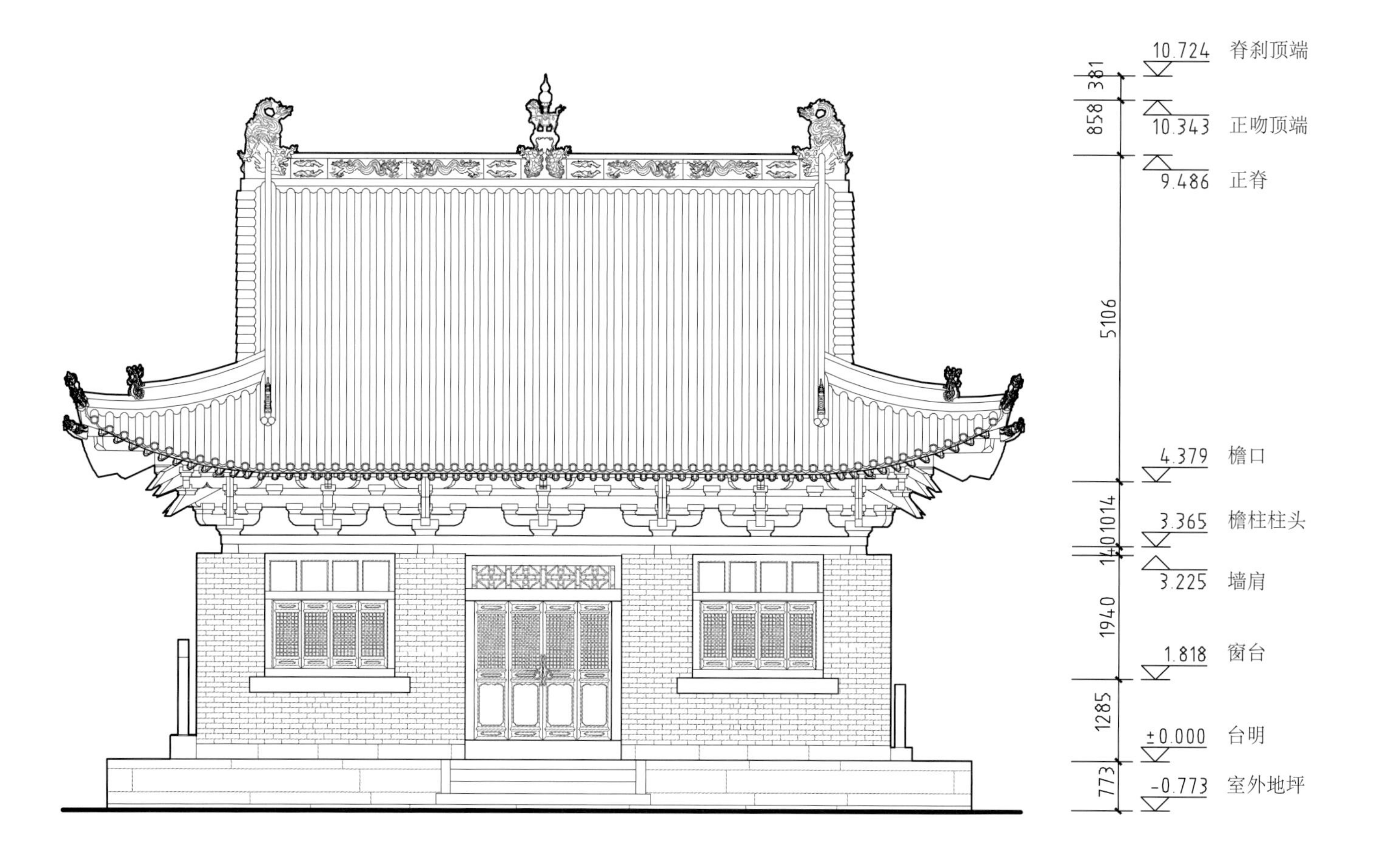

游仙寺毗卢殿正立面图

Front elevation of Youxian Monastery's Pilu Hall

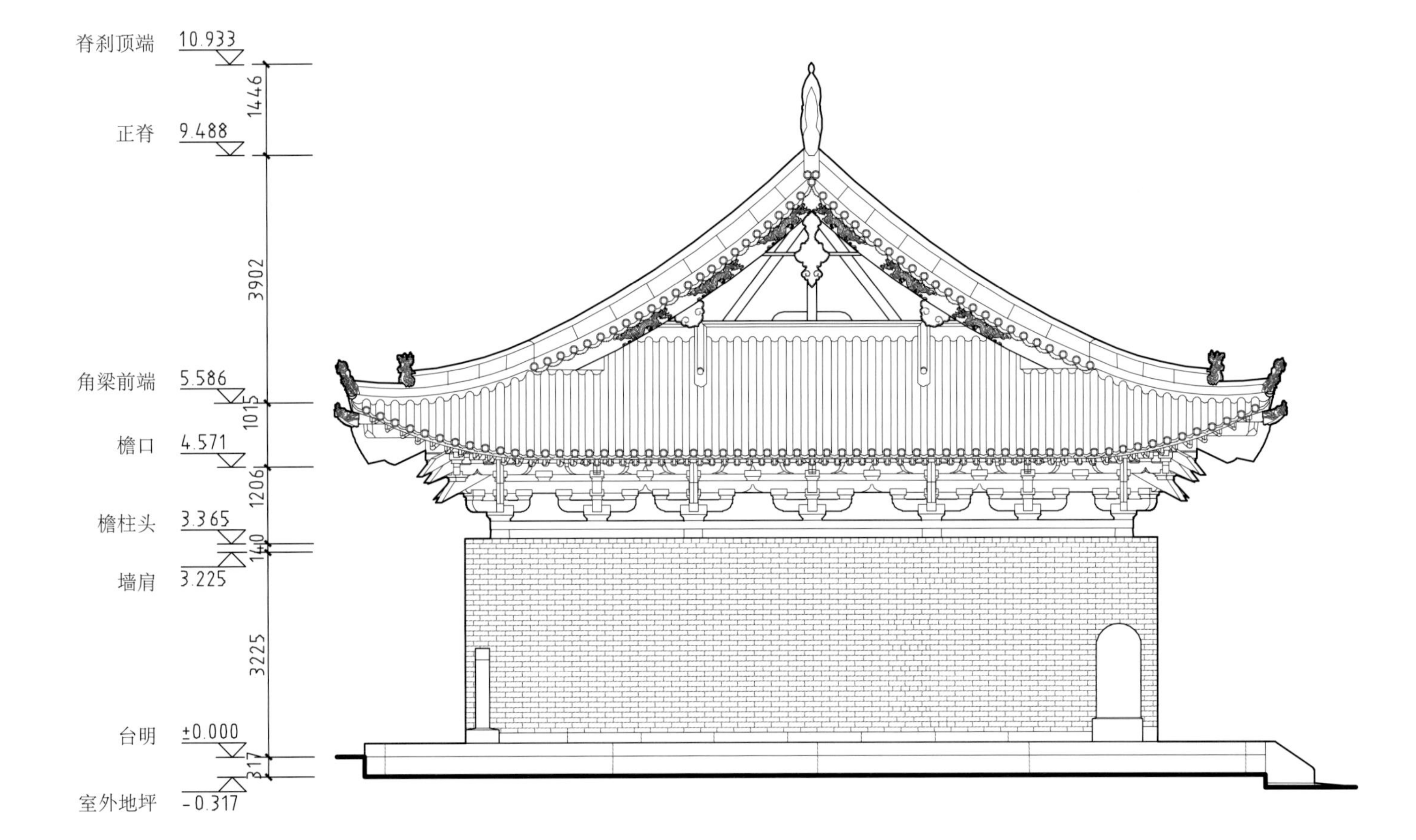

游仙寺毗卢殿侧立面图
Side elevation of Youxian Monastery's Pilu Hall

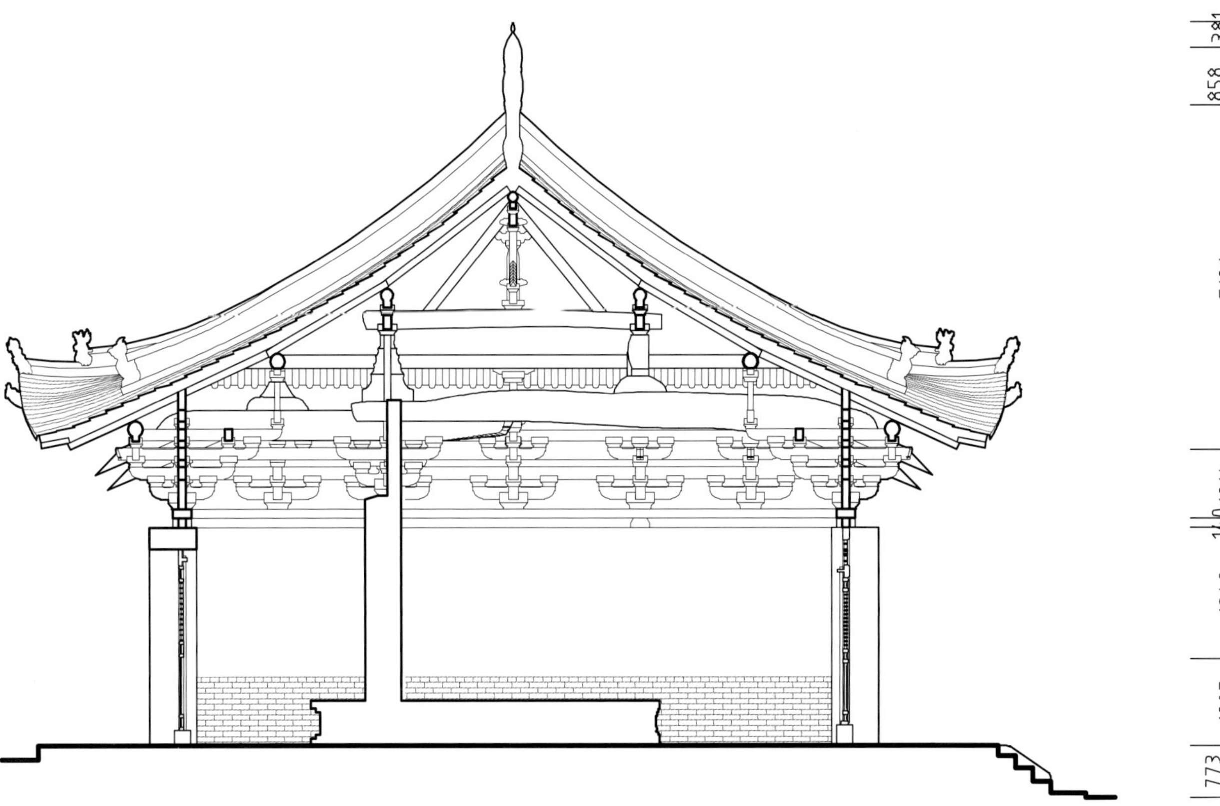

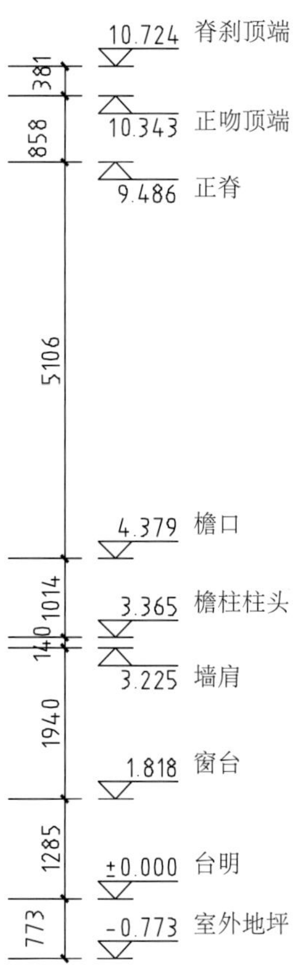

游仙寺毗卢殿横剖面图

Cross-section of Youxian Monastery's Pilu Hall

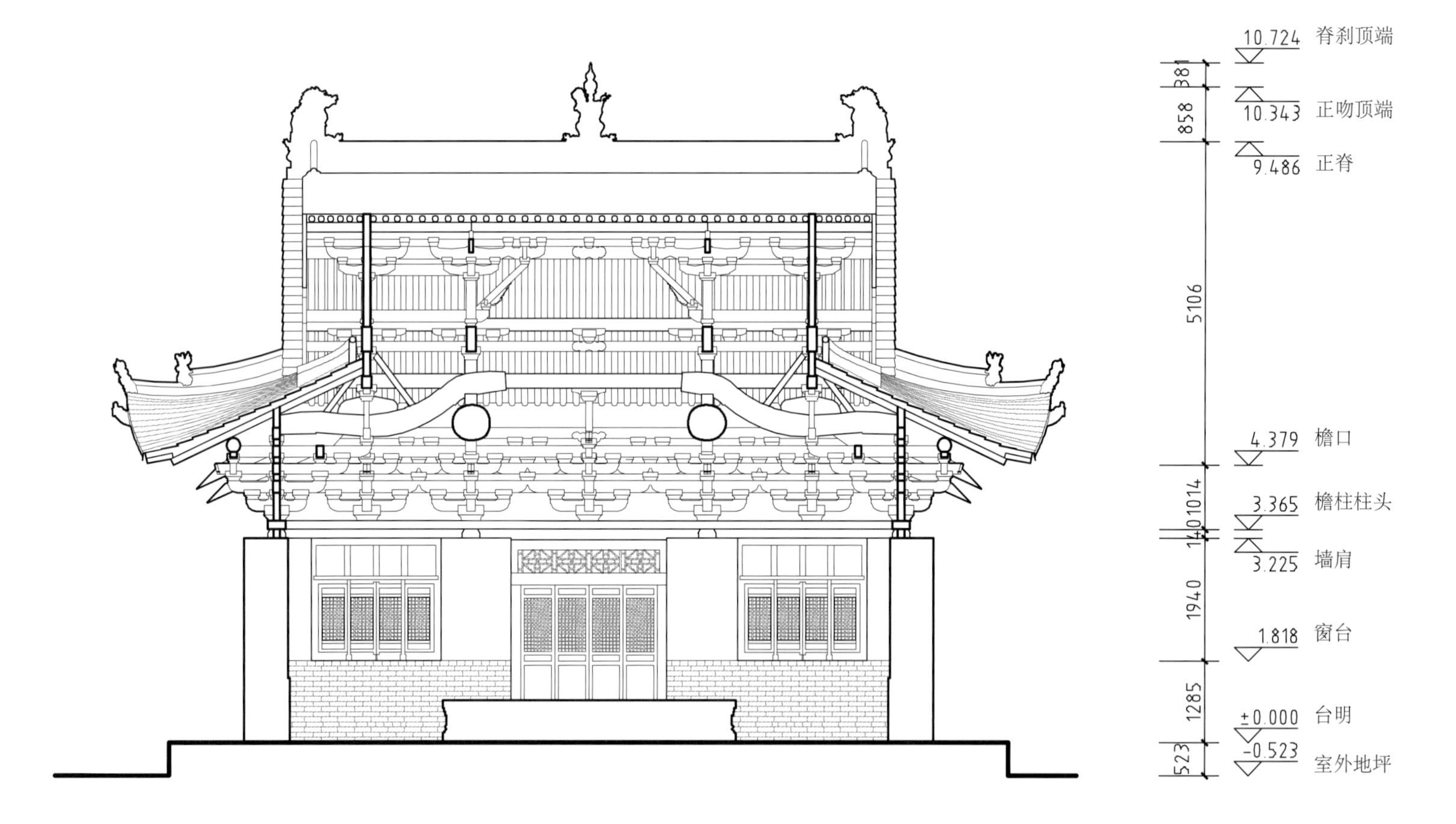

游仙寺毗卢殿纵剖面图

Longitudinal section of Youxian Monastery's Pilu Hall

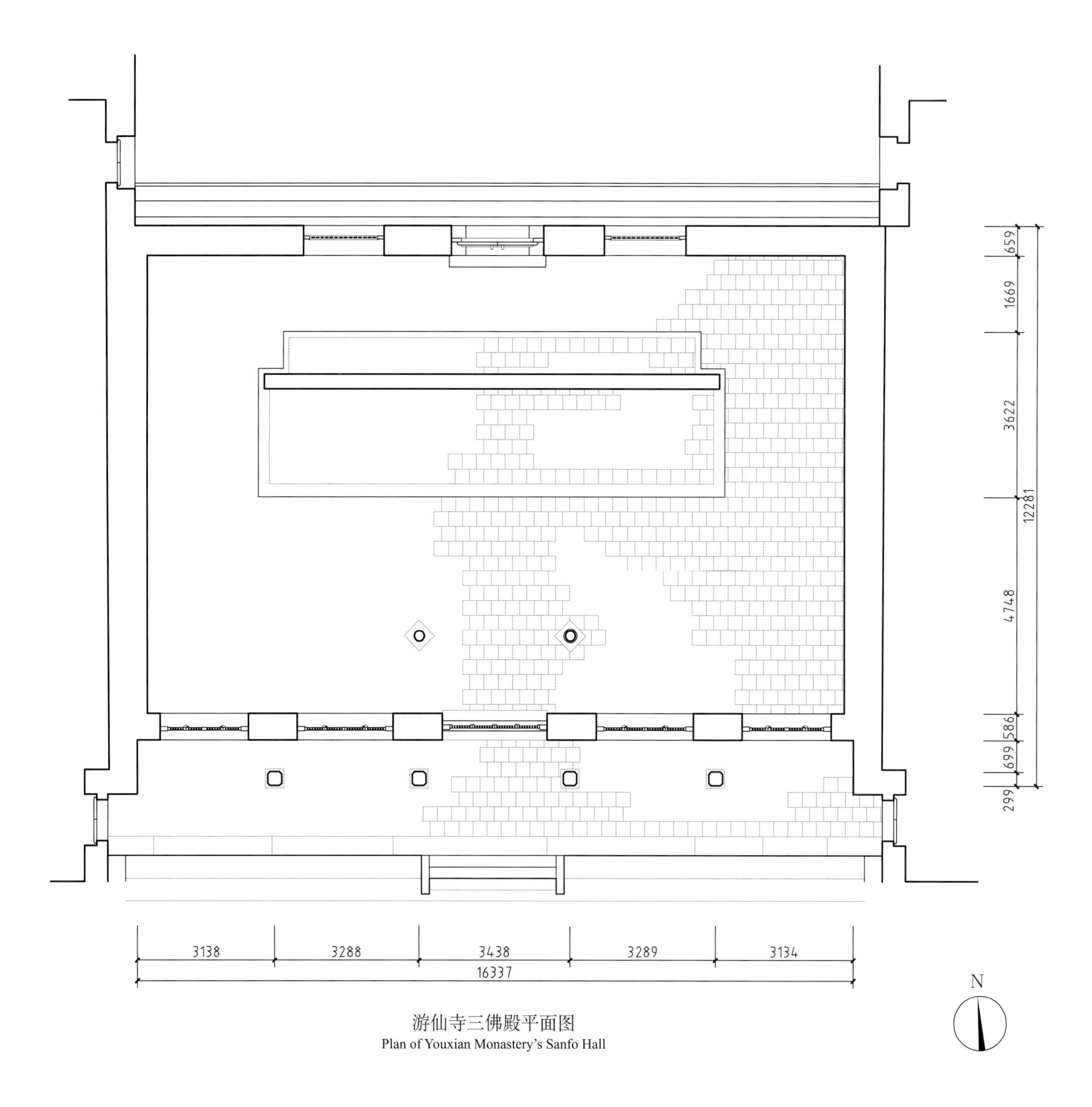

游仙寺三佛殿平面图

Plan of Youxian Monastery's Sanfo Hall

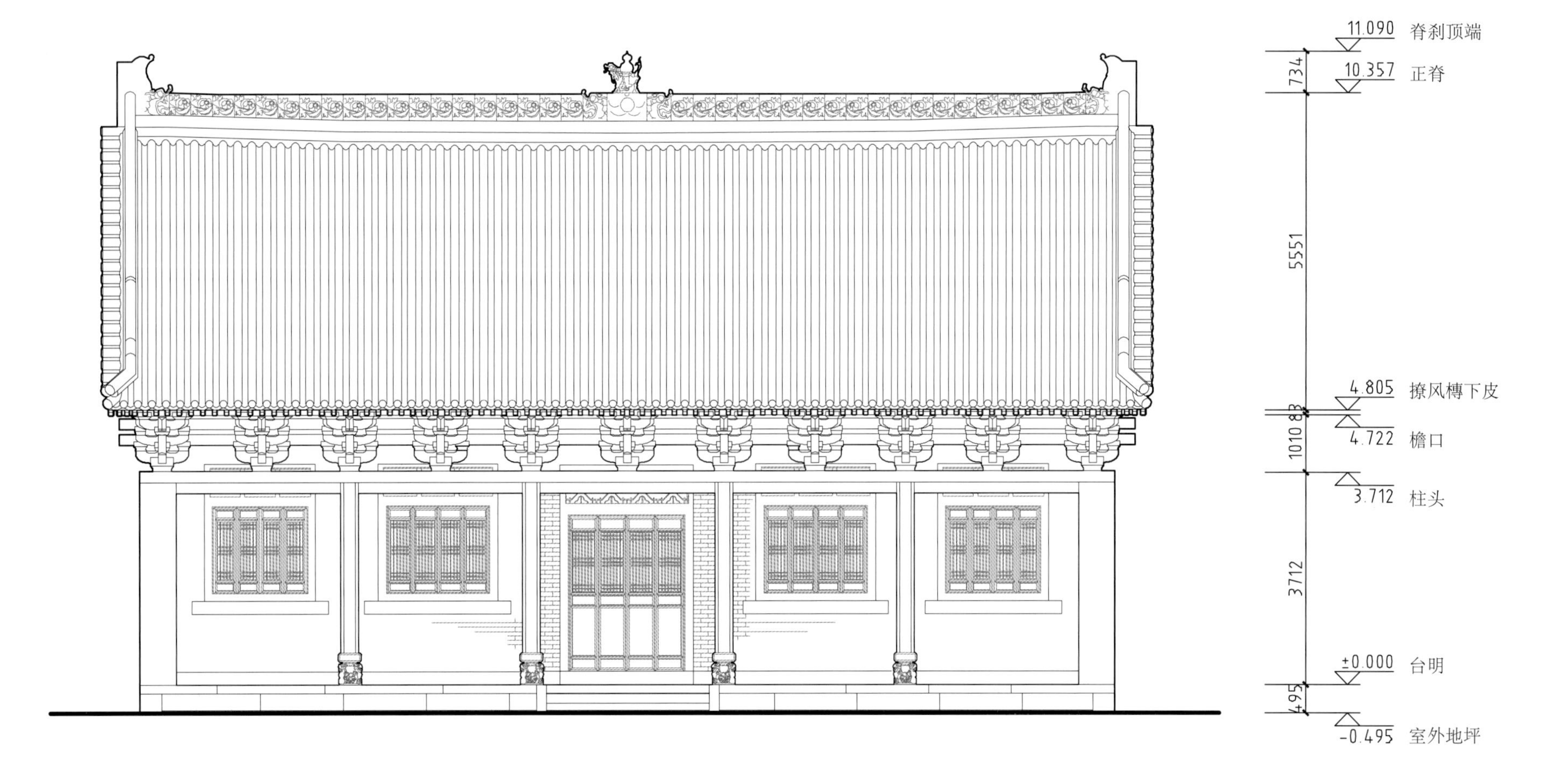

游仙寺三佛殿正立面图
Front elevation of Youxian Monastery's Sanfo Hall

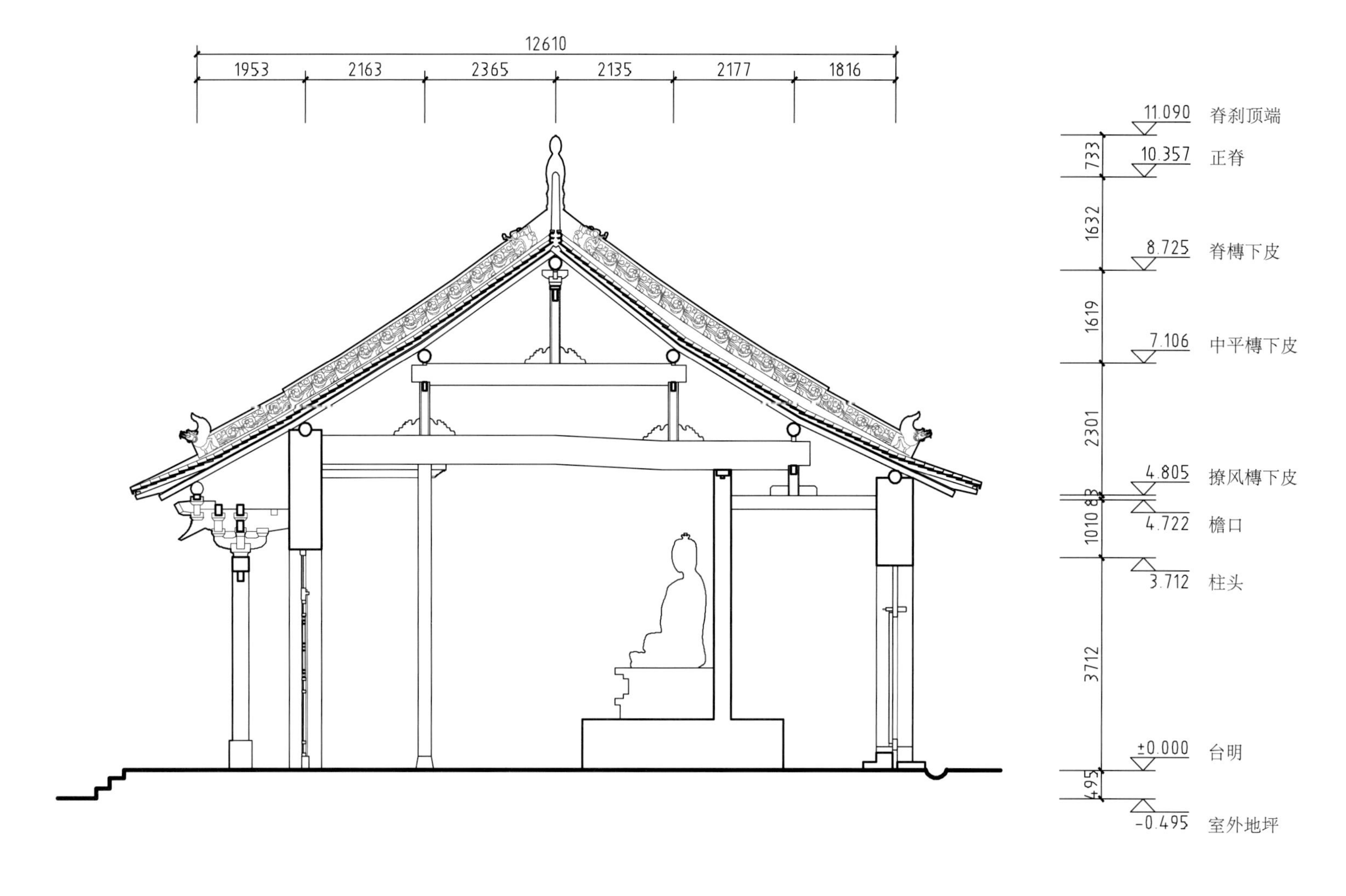

游仙寺三佛殿横剖面图

Cross-section of Youxian Monastery's Sanfo Hall

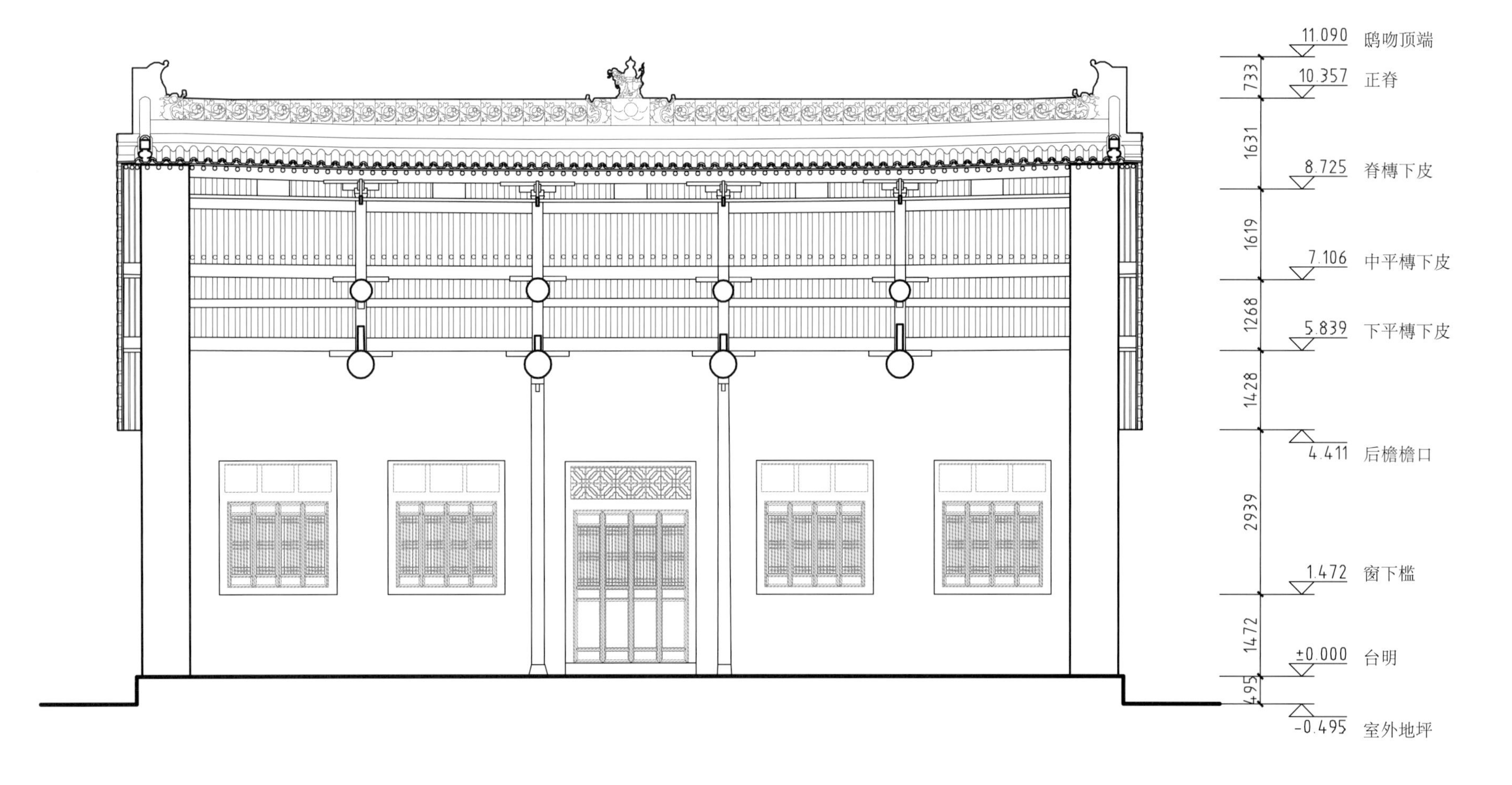

游仙寺三佛殿纵剖面图

Longitudinal section of Youxian Monastery's Sanfo Hall

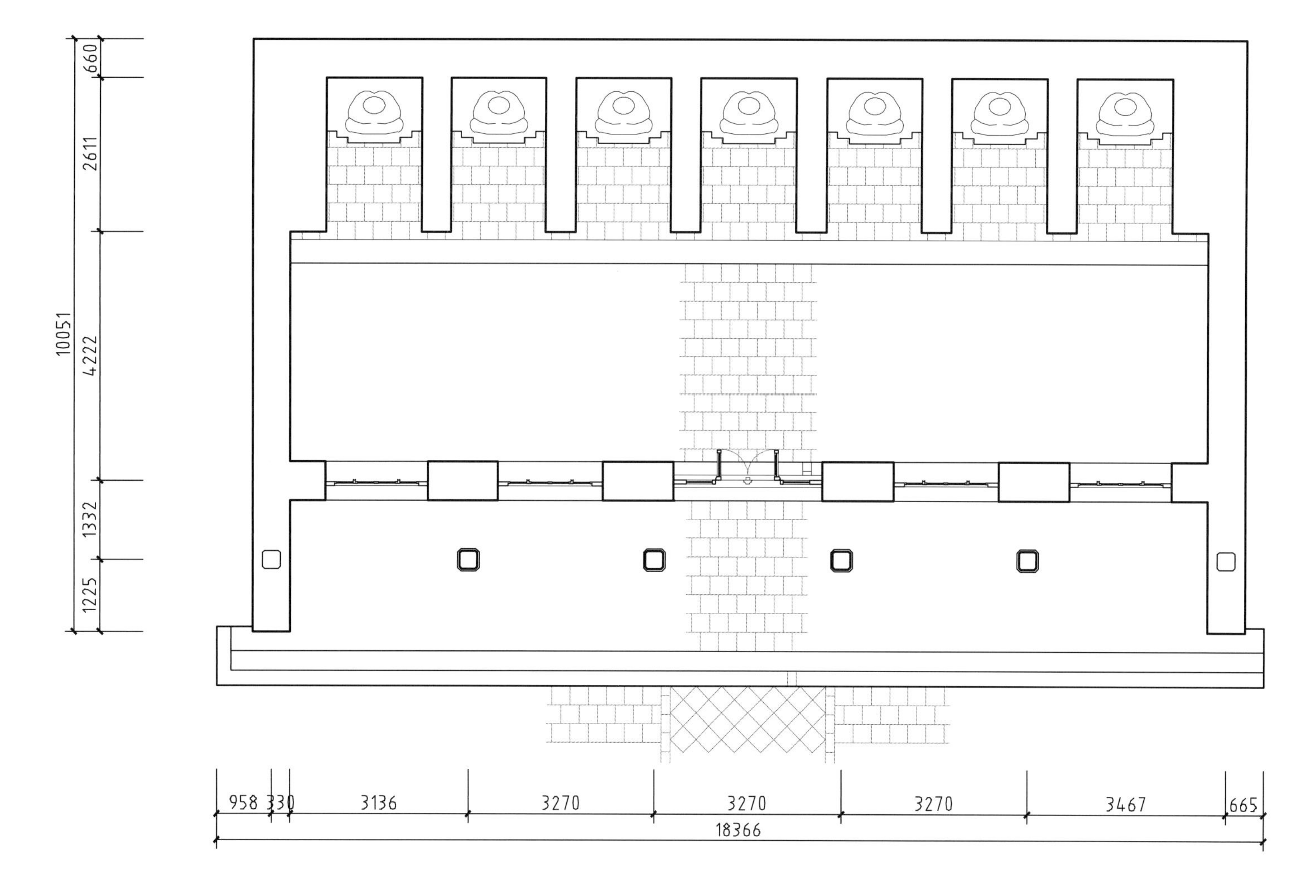

游仙寺七佛殿平面图

Plan of Youxian Monastery's Qifo Hall

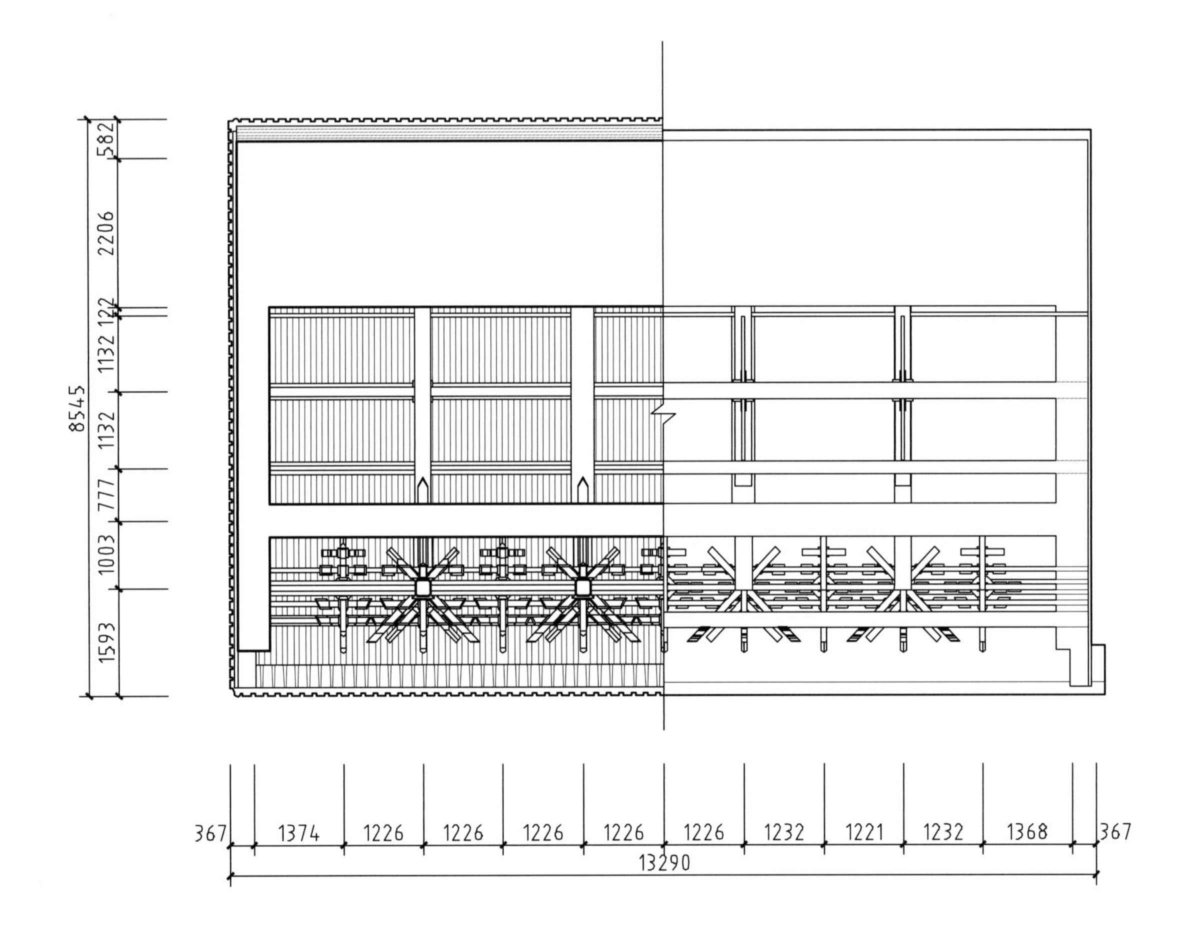

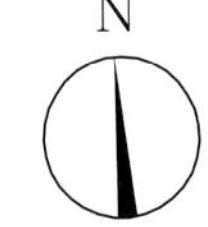

游仙寺七佛殿梁架俯仰视平面图
Plan of Youxian Monastery's Qifo Hall's framework as seen from above & below

游仙寺七佛殿正立面图
Front elevation of Youxian Monastery's Qifo Hall

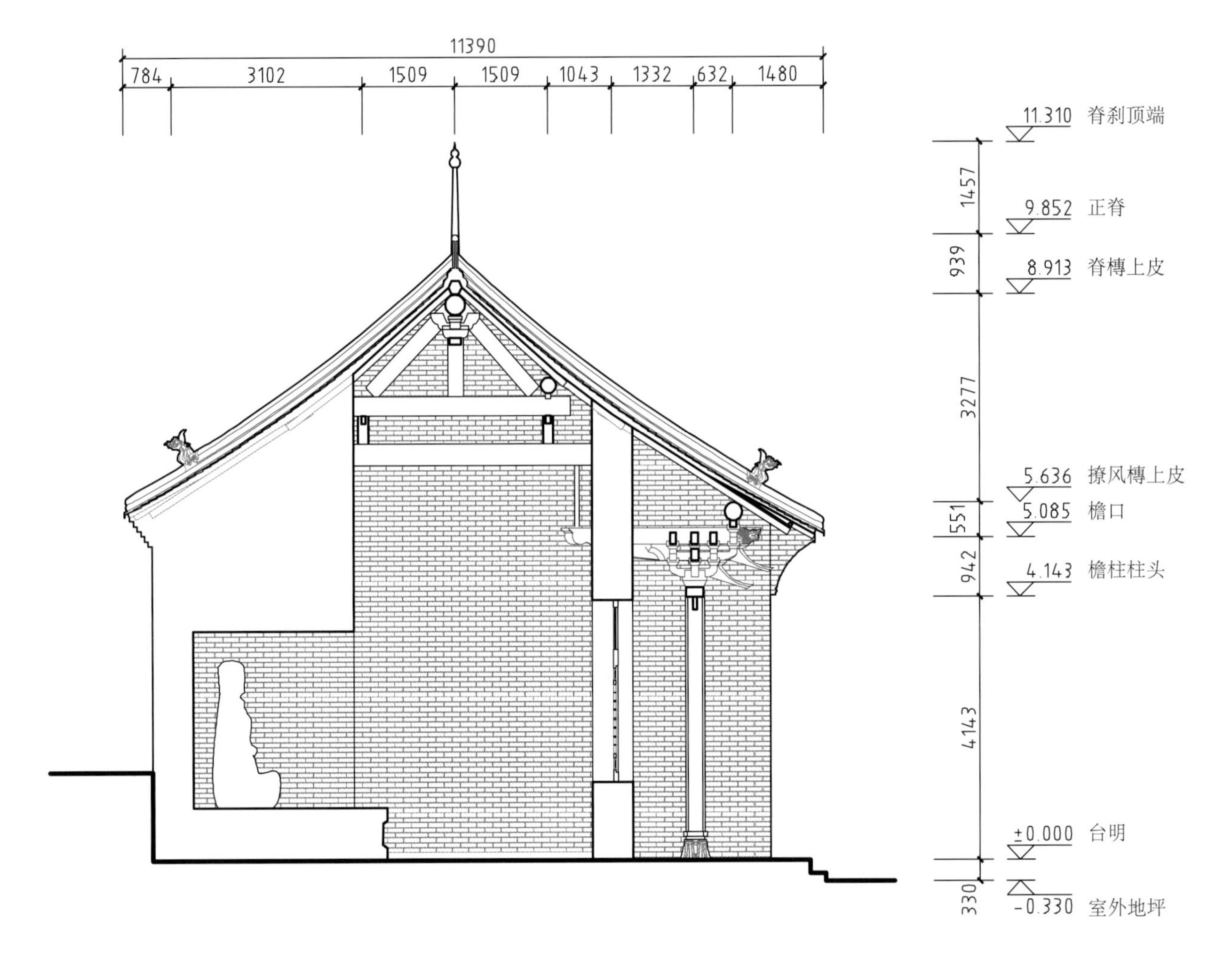

游仙寺七佛殿横剖面图
Cross-section of Youxian Monastery's Qifo Hall

开化寺
Kaihua Monastery

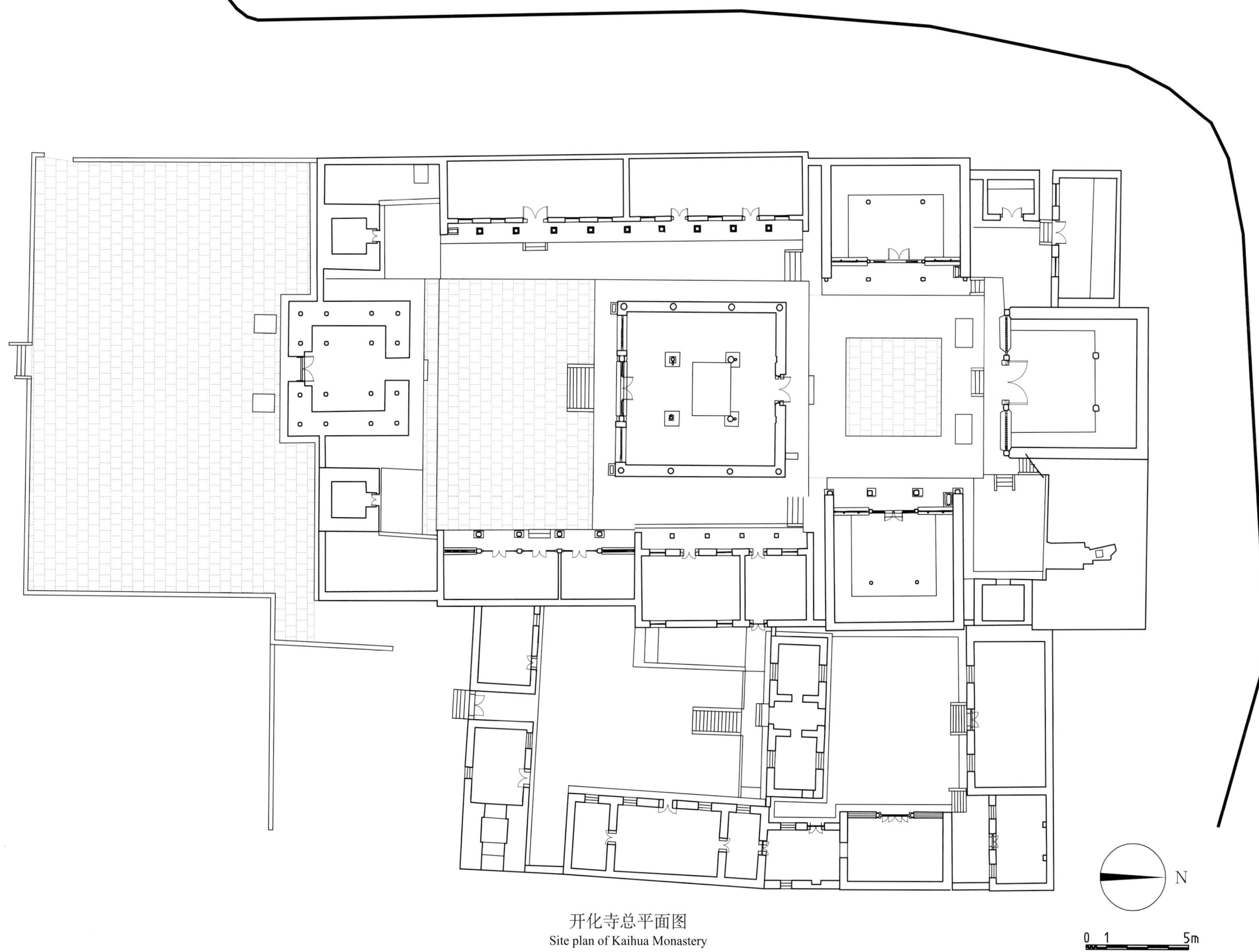

开化寺总平面图
Site plan of Kaihua Monastery

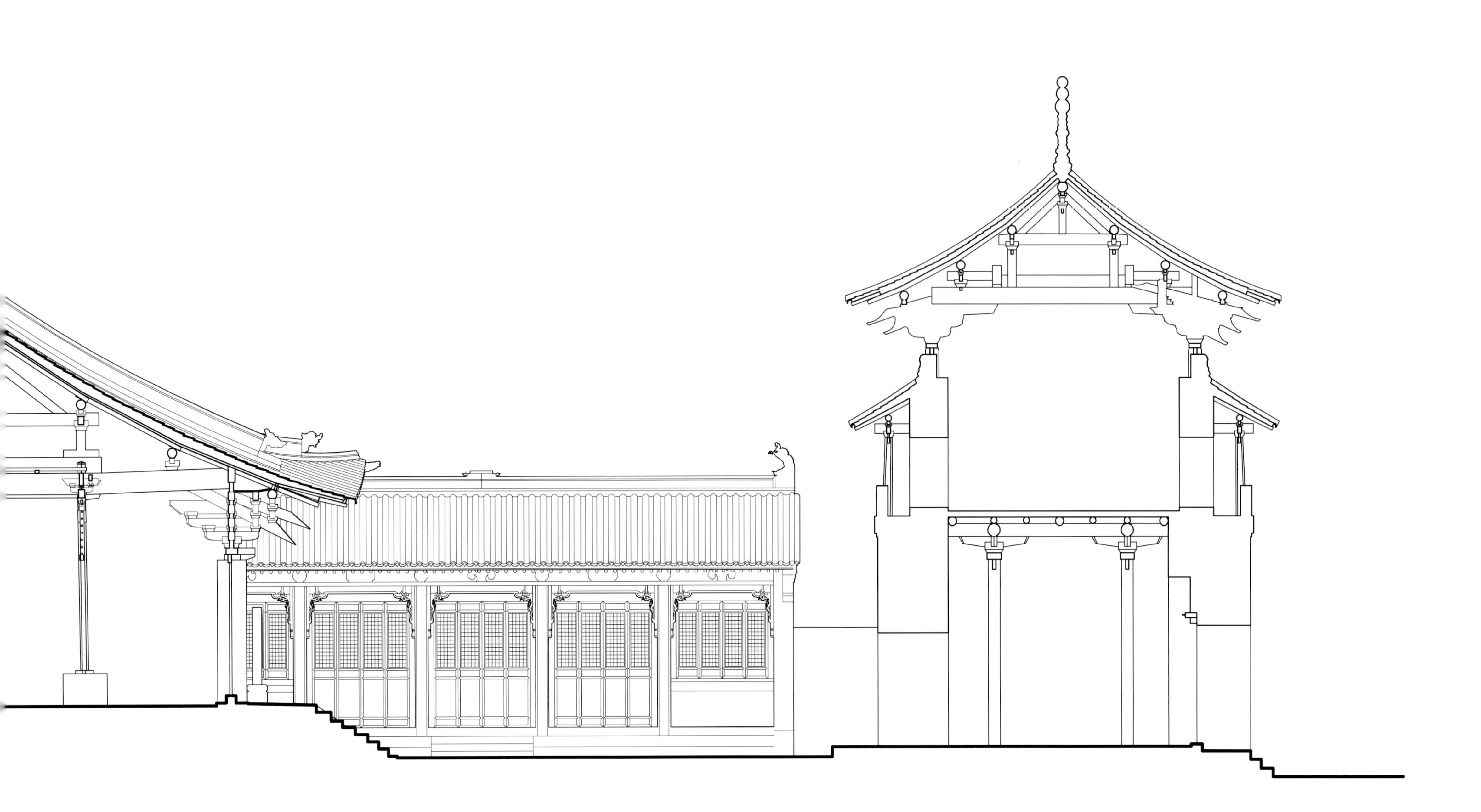
0 1 5m

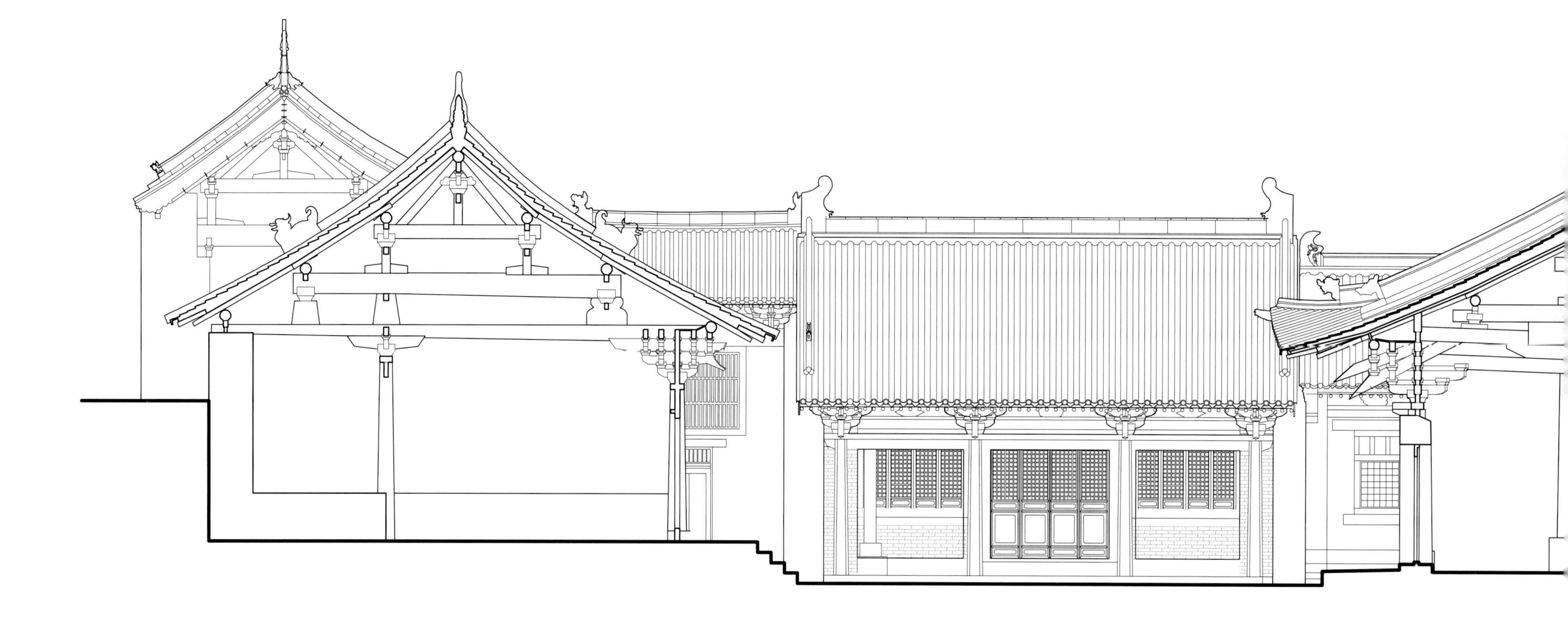

开化寺总纵剖面图
Longitudinal site section of Kaihua Monastery

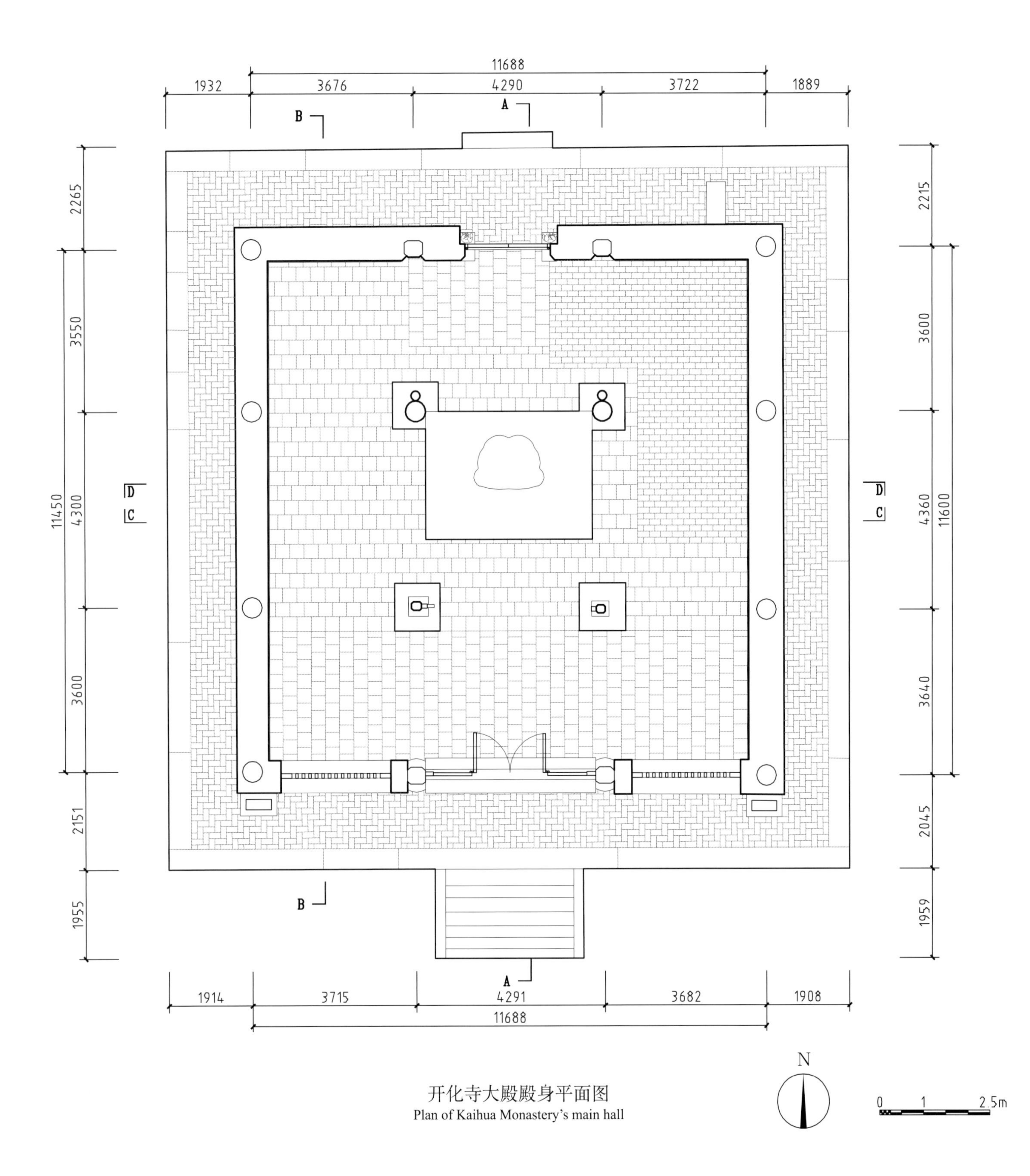

开化寺大殿殿身平面图
Plan of Kaihua Monastery's main hall

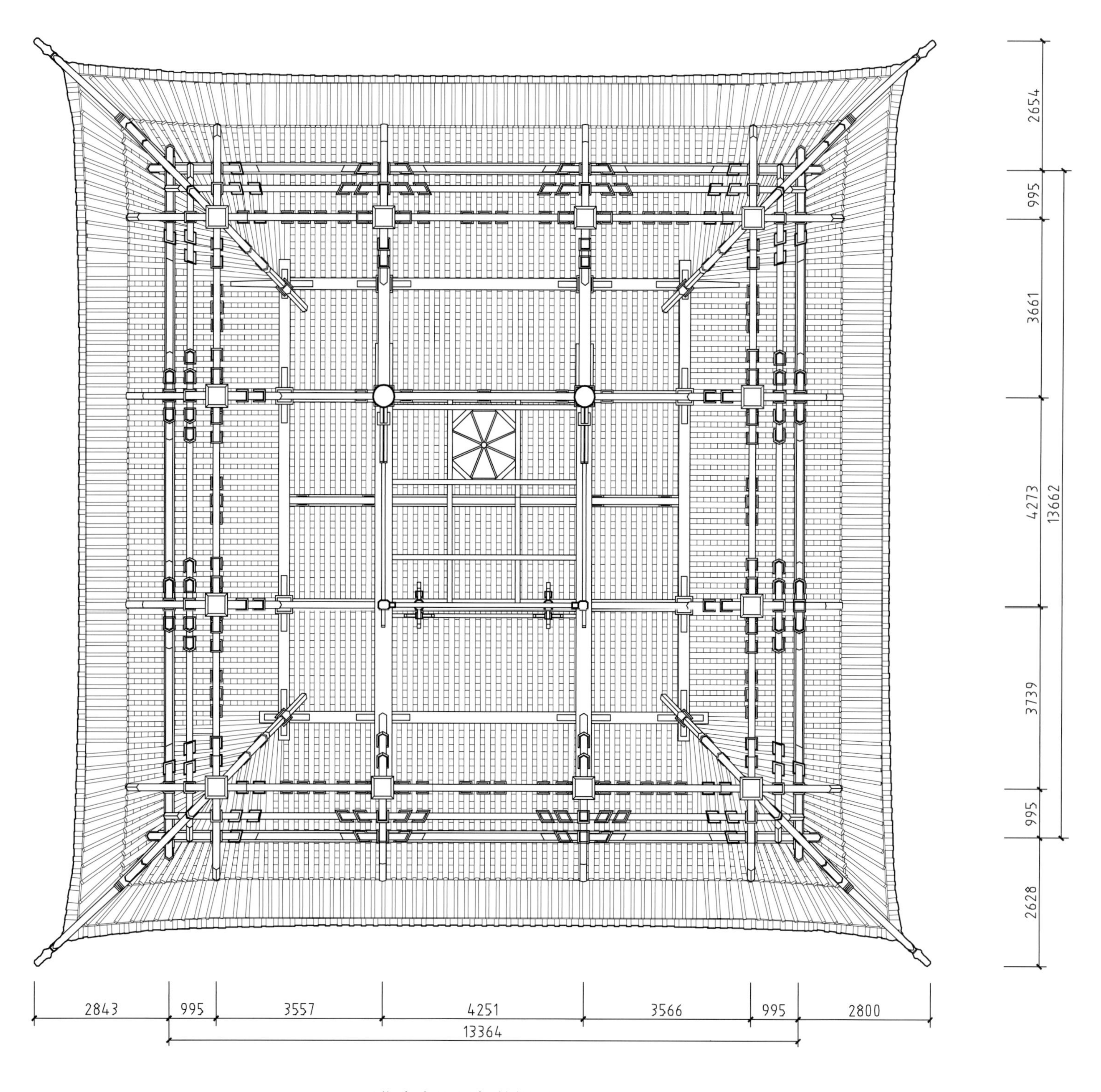

开化寺大殿梁架仰视平面图

Plan of Kaihua Monastery's main hall's framework as seen from below

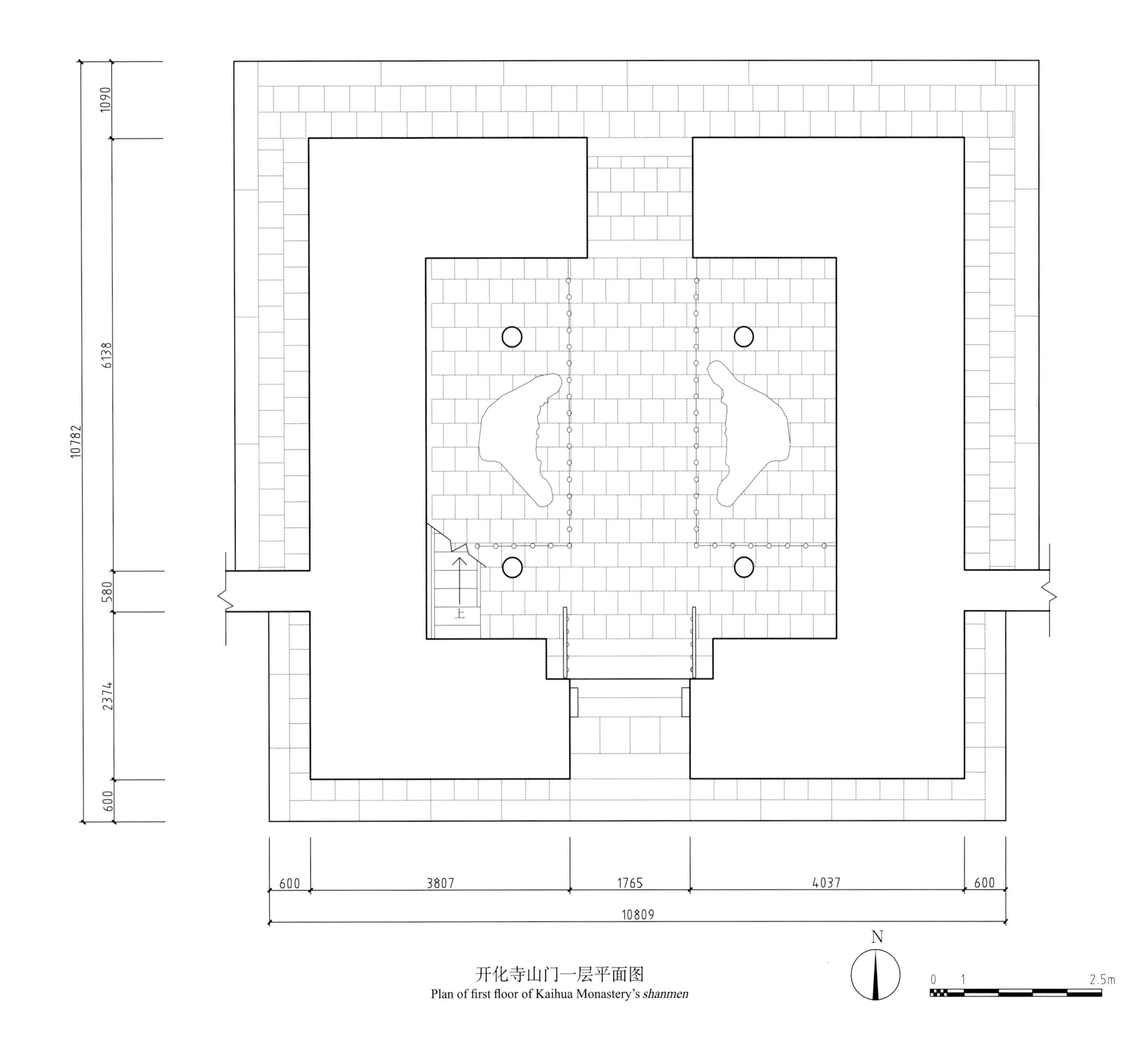

开化寺山门一层平面图

Plan of first floor of Kaihua Monastery's *shanmen*

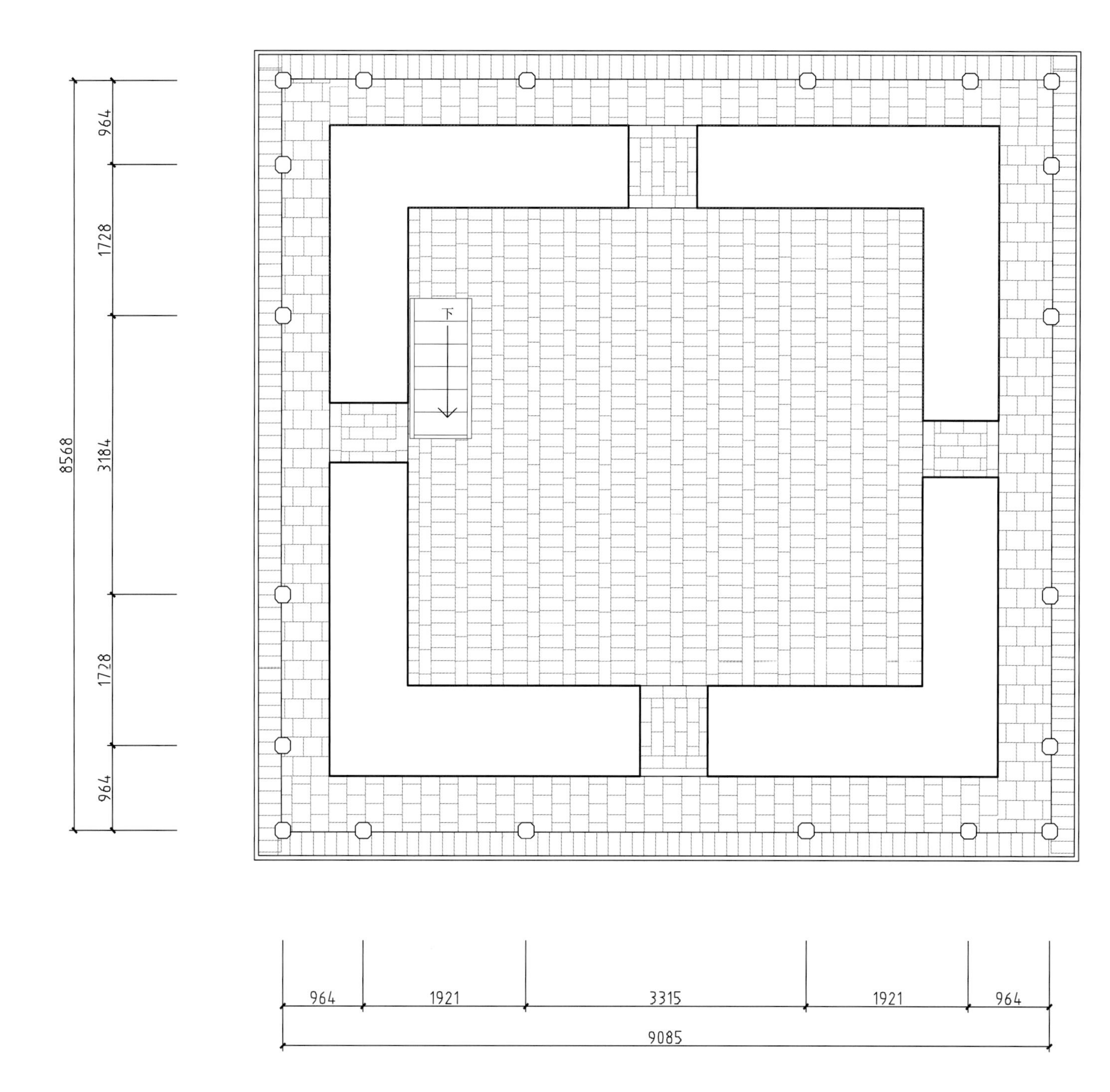

0 1 2.5m

开化寺山门二层平面图

Plan of second floor of Kaihua Monastery's *shanmen*

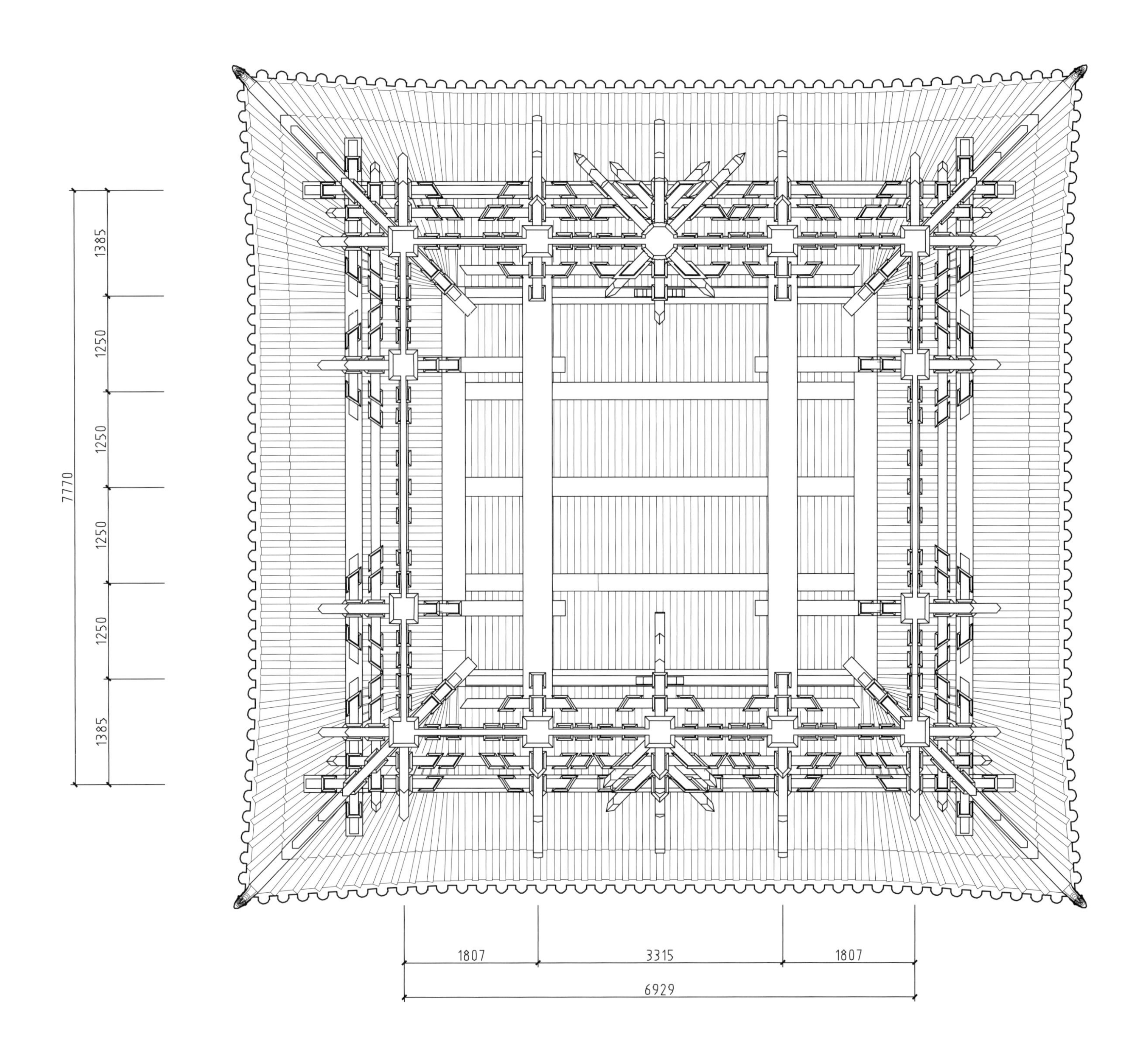

开化寺山门二层仰视平面图

Plan of second floor of Kaihua Monastery's *shanmen* as seen from below

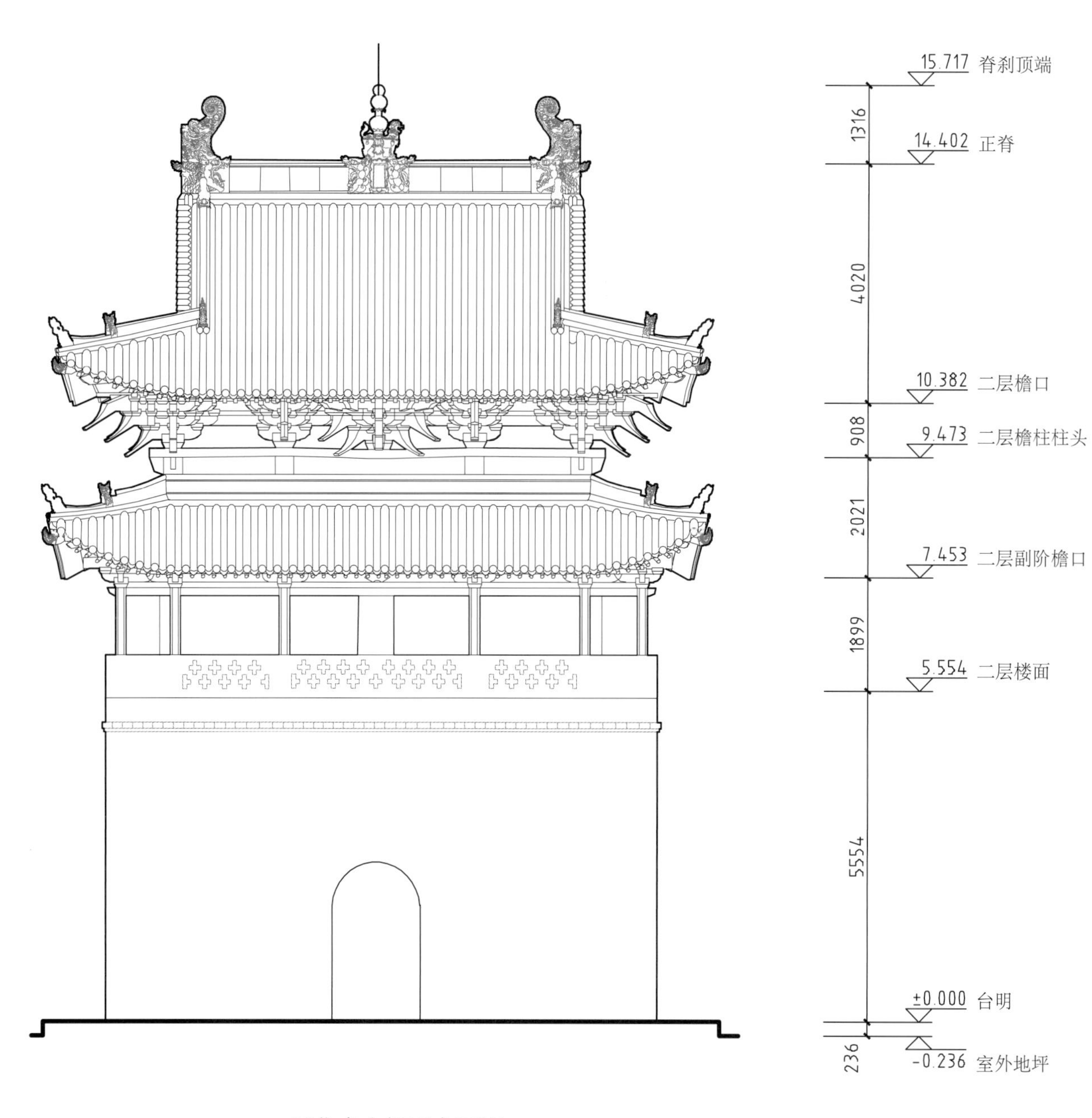

开化寺山门正立面图

Front elevation of Kaihua Monastery's *shanmen*

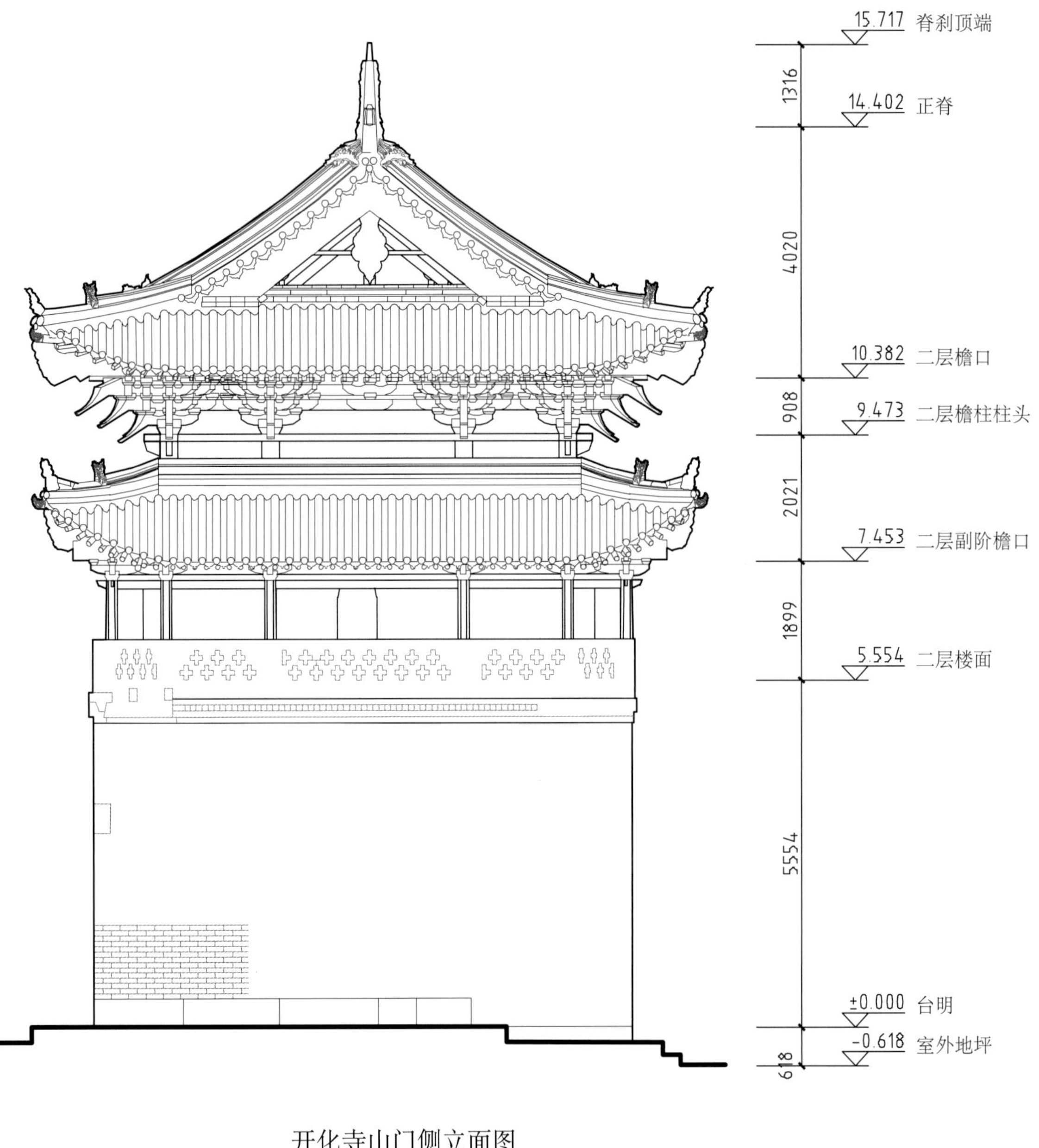

开化寺山门侧立面图

Side elevation of Kaihua Monastery's *shanmen*

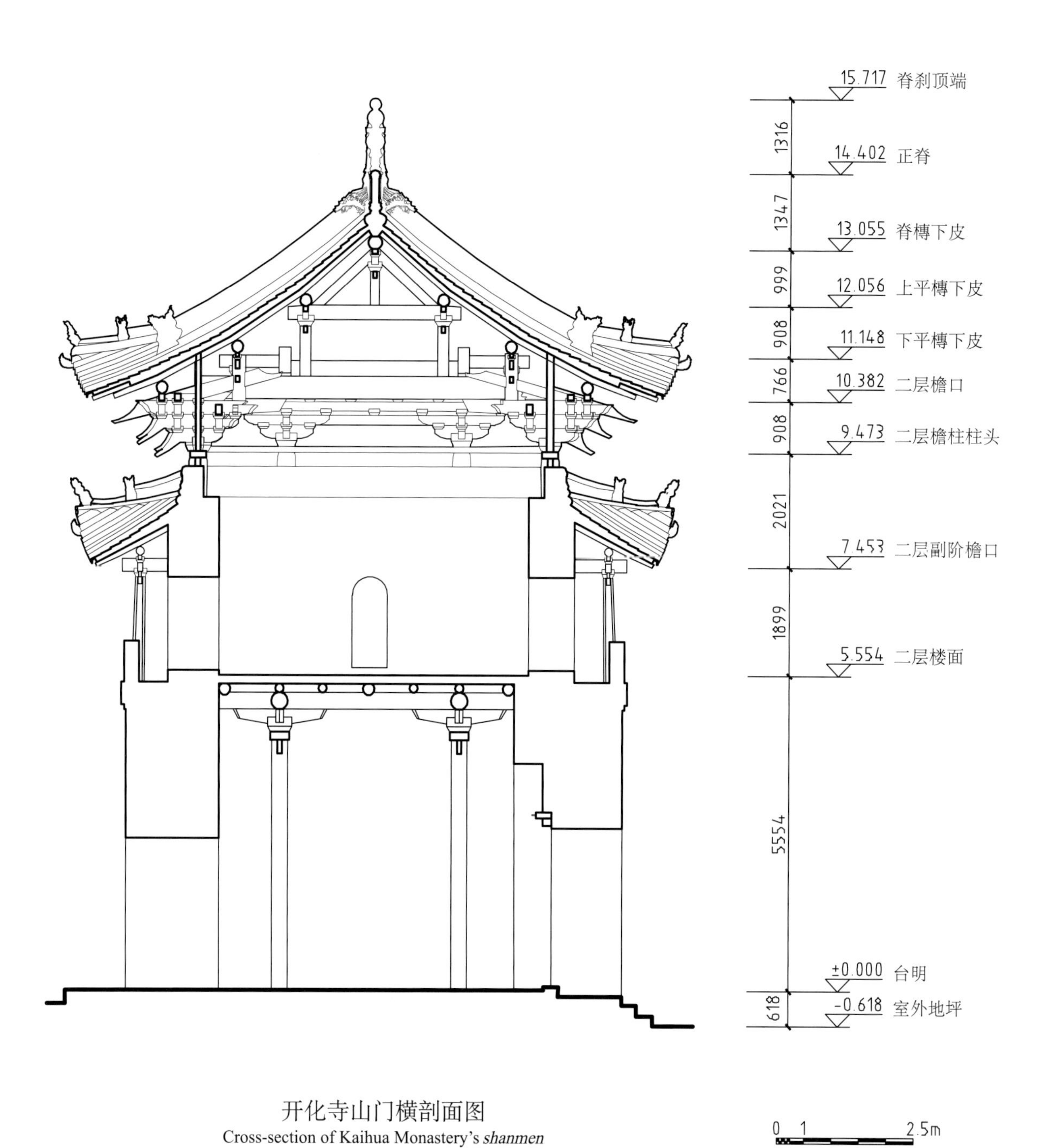

开化寺山门横剖面图

Cross-section of Kaihua Monastery's *shanmen*

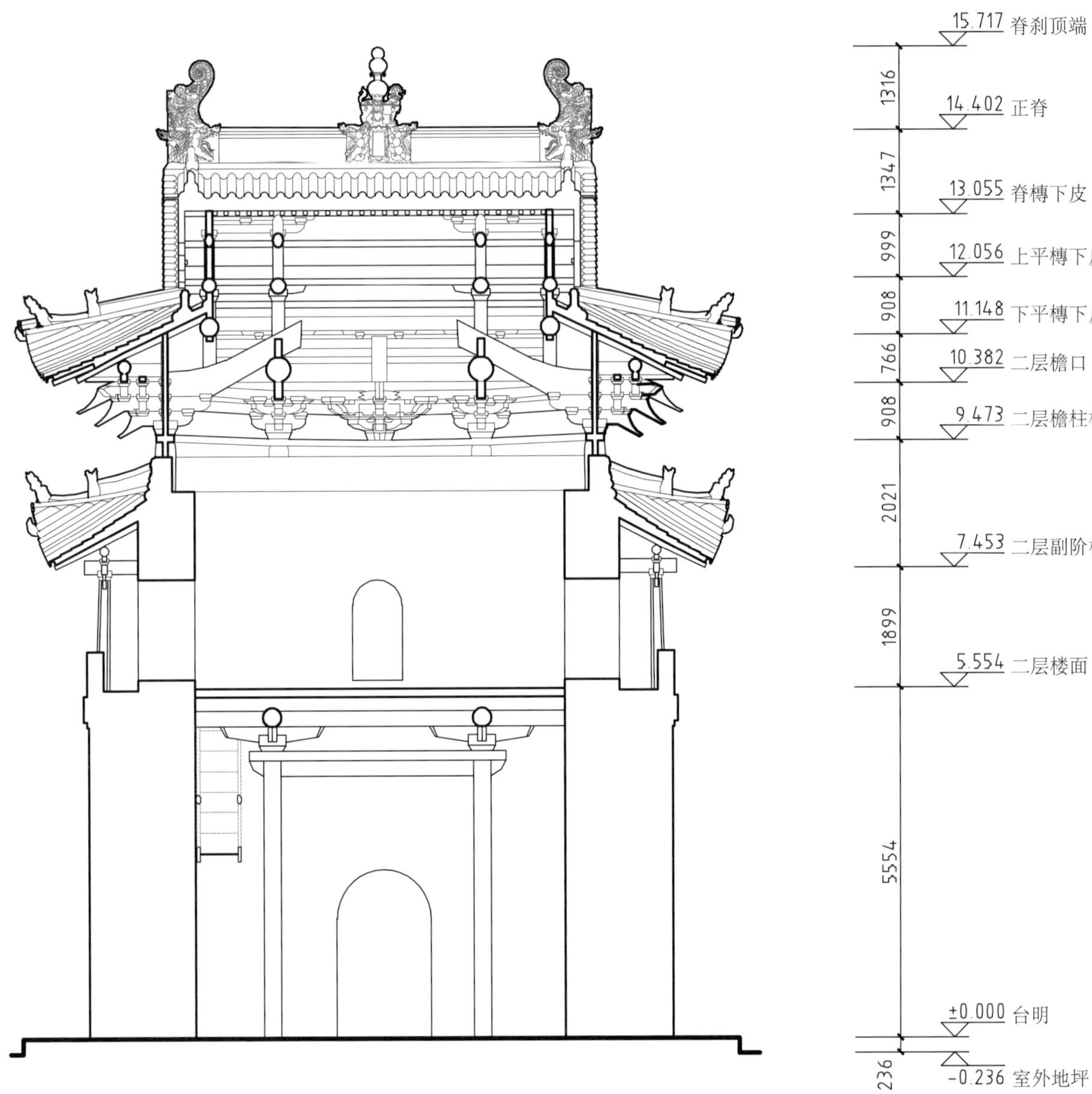

开化寺山门纵剖面图

Longitudinal section of Kaihua Monastery's *shanmen*

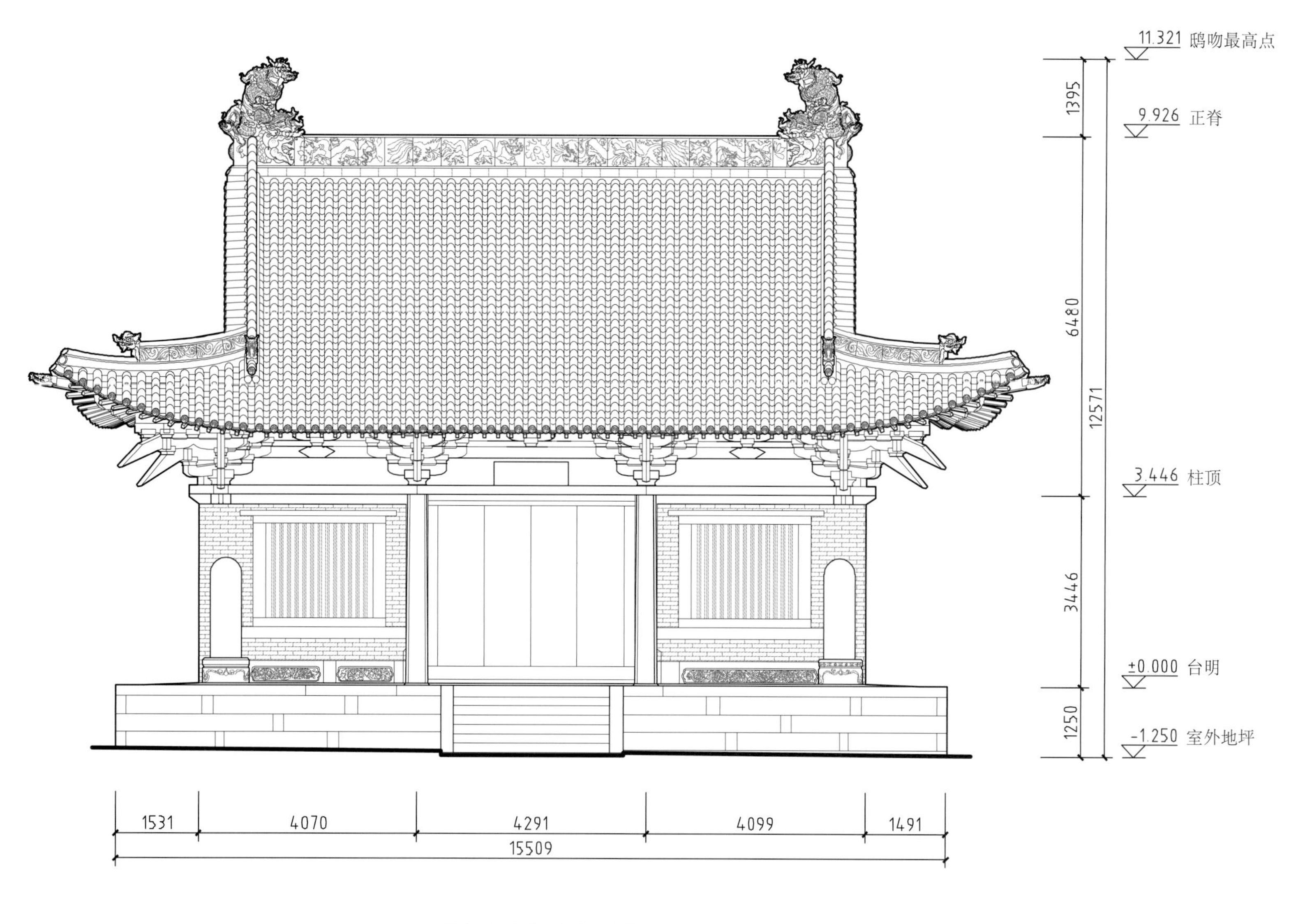

开化寺大殿南立面图
South elevation of Kaihua Monastery's main hall

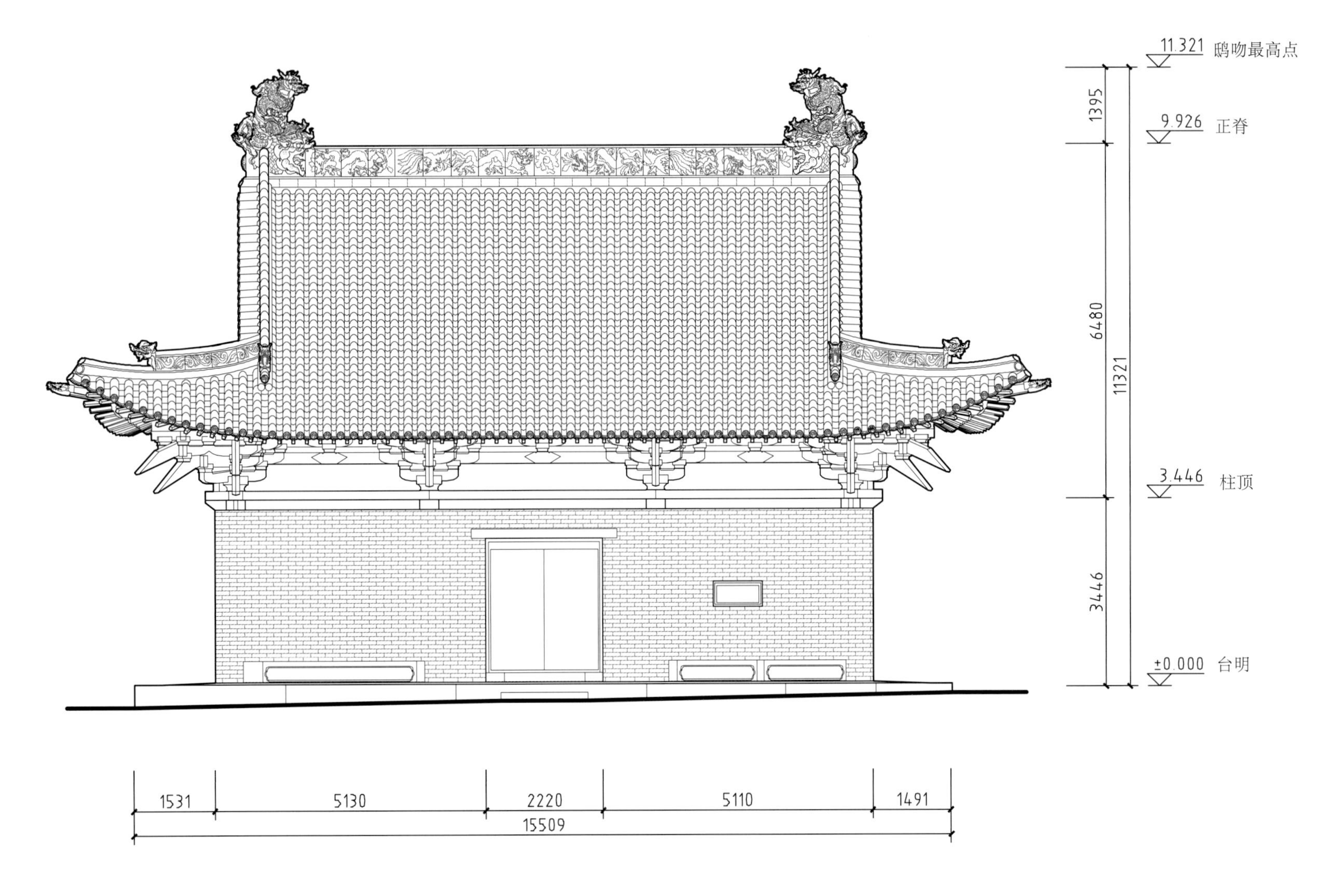

开化寺大殿北立面图

North elevation of Kaihua Monastery's main hall

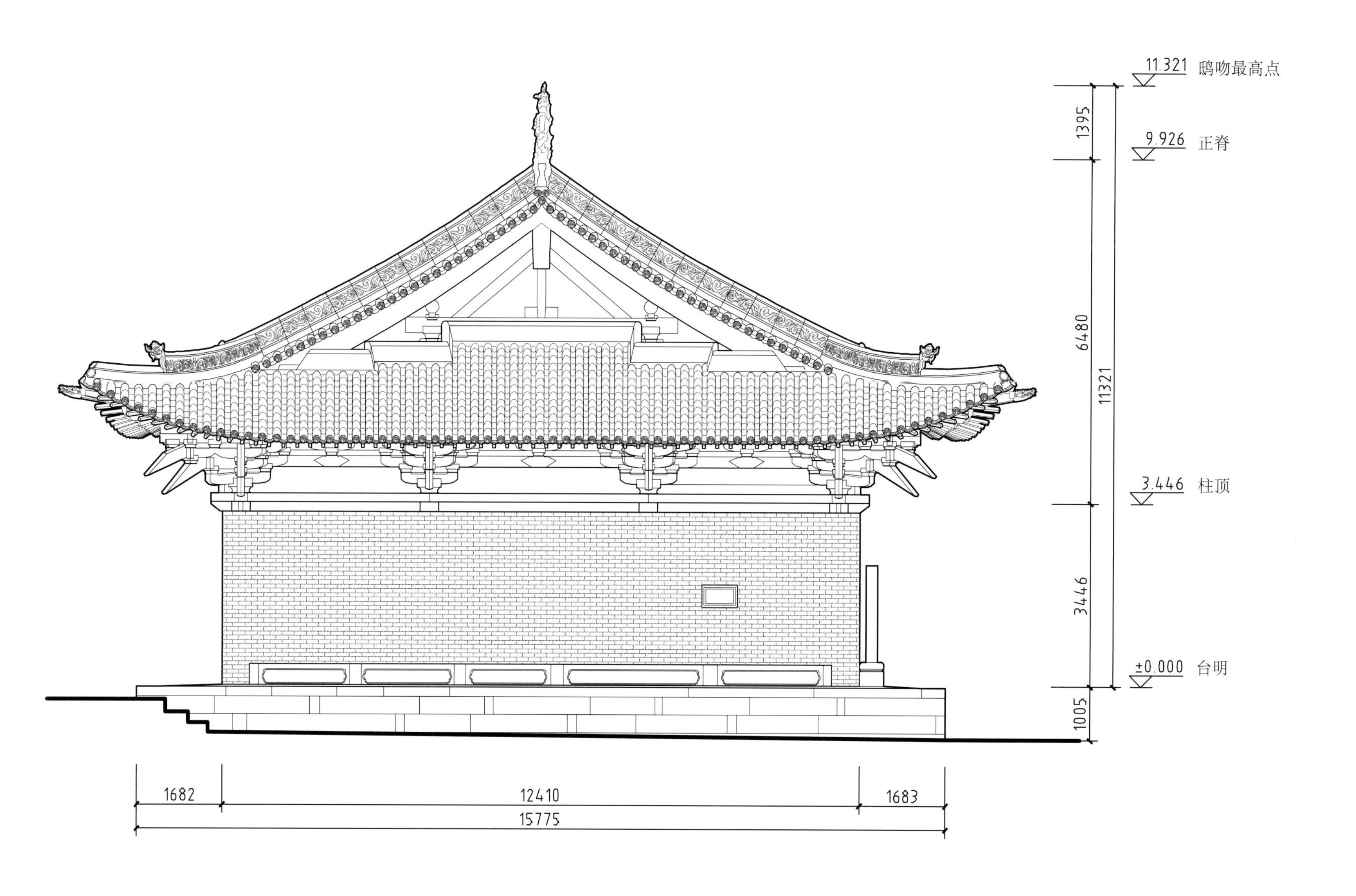

开化寺大殿西立面图

West elevation of Kaihua Monastery's main hall

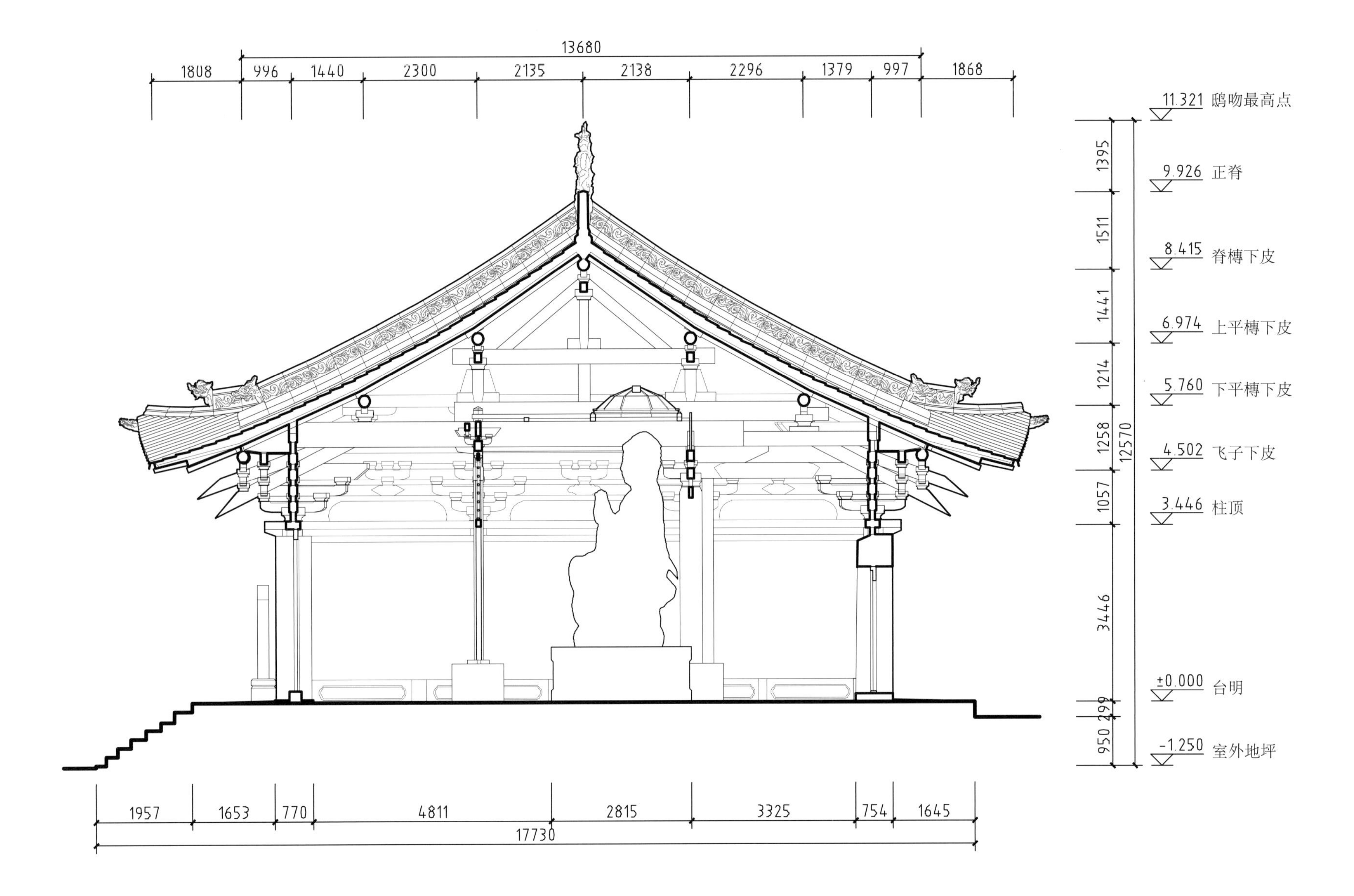

开化寺大殿 A-A 剖面图

Section A-A of Kaihua Monastery's main hall

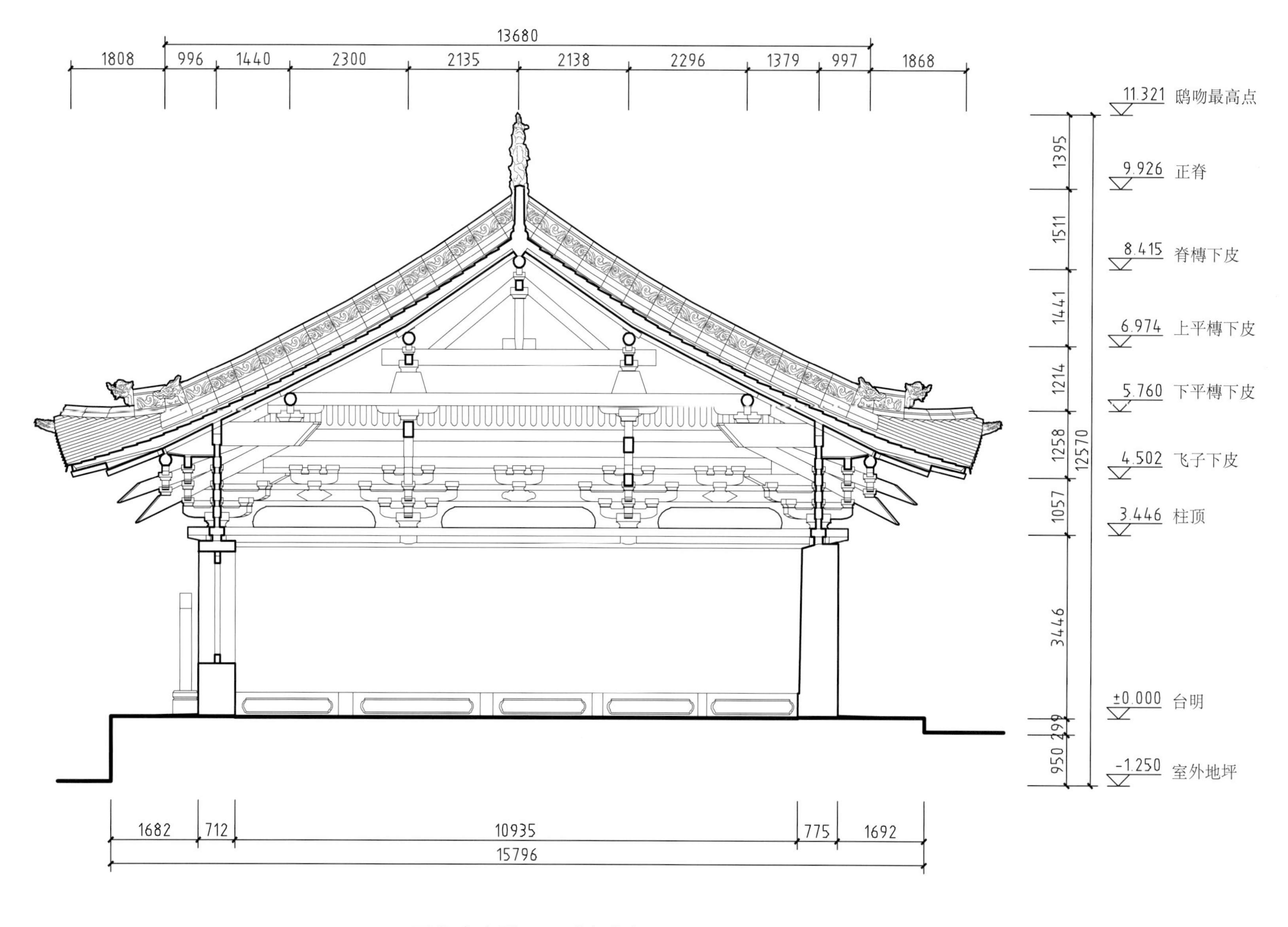

开化寺大殿 B-B 剖面图

Section B-B of Kaihua Monastery's main hall

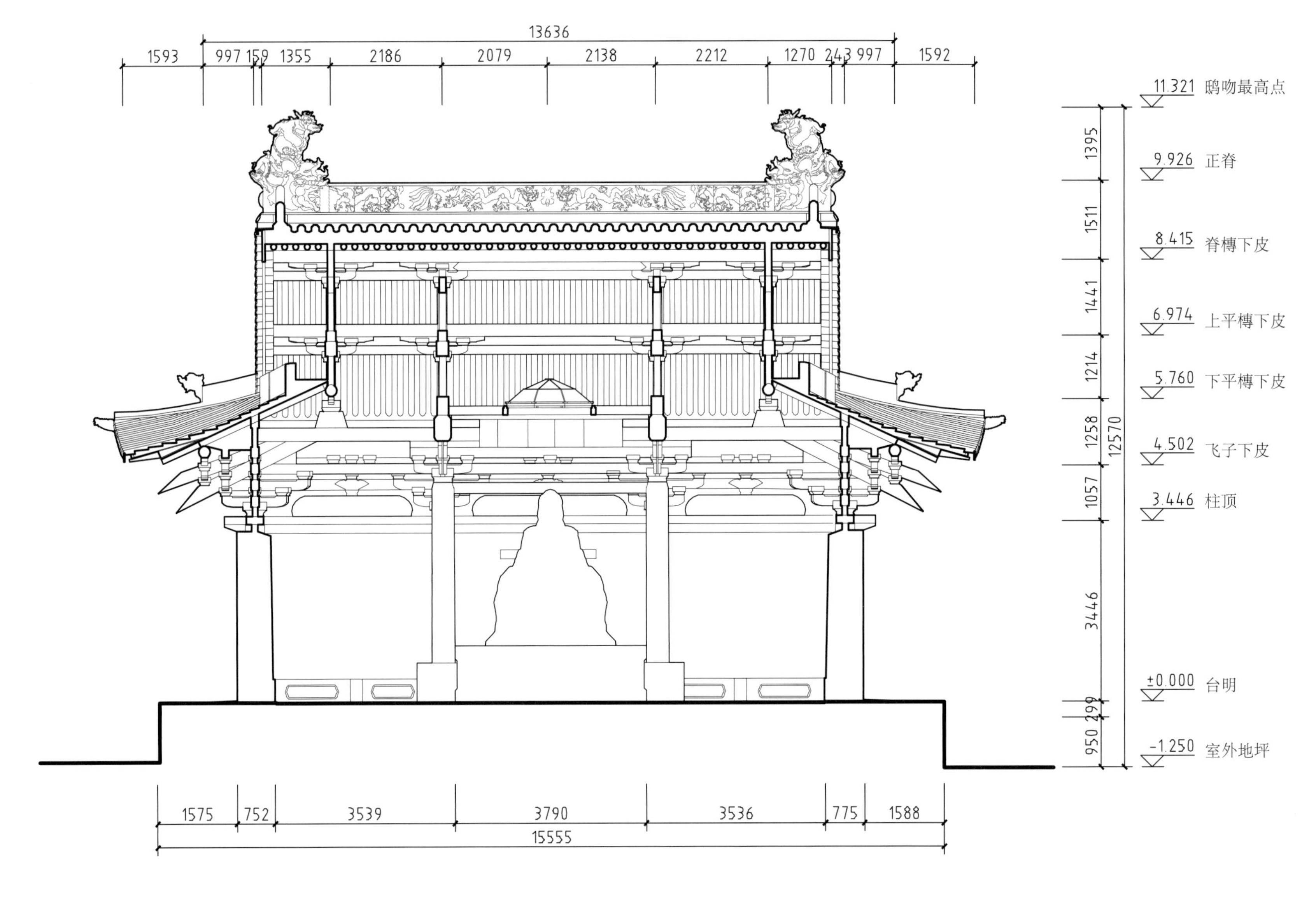

开化寺大殿 C–C 剖面图

Longitudinal section C-C of Kaihua Monastery's main hall

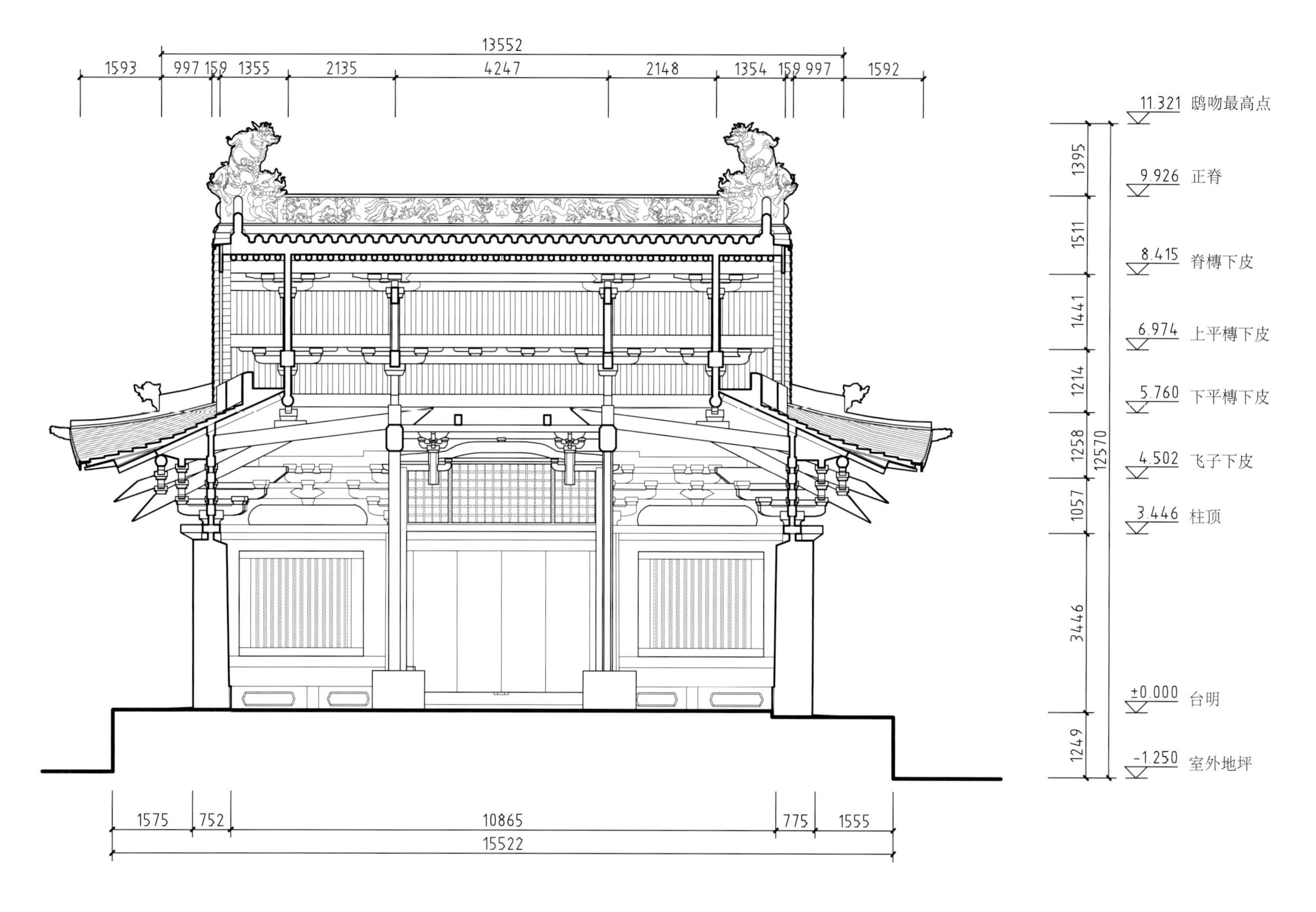

开化寺大殿 D-D 剖面图

Longitudinal section D-D of Kaihua Monastery's main hall

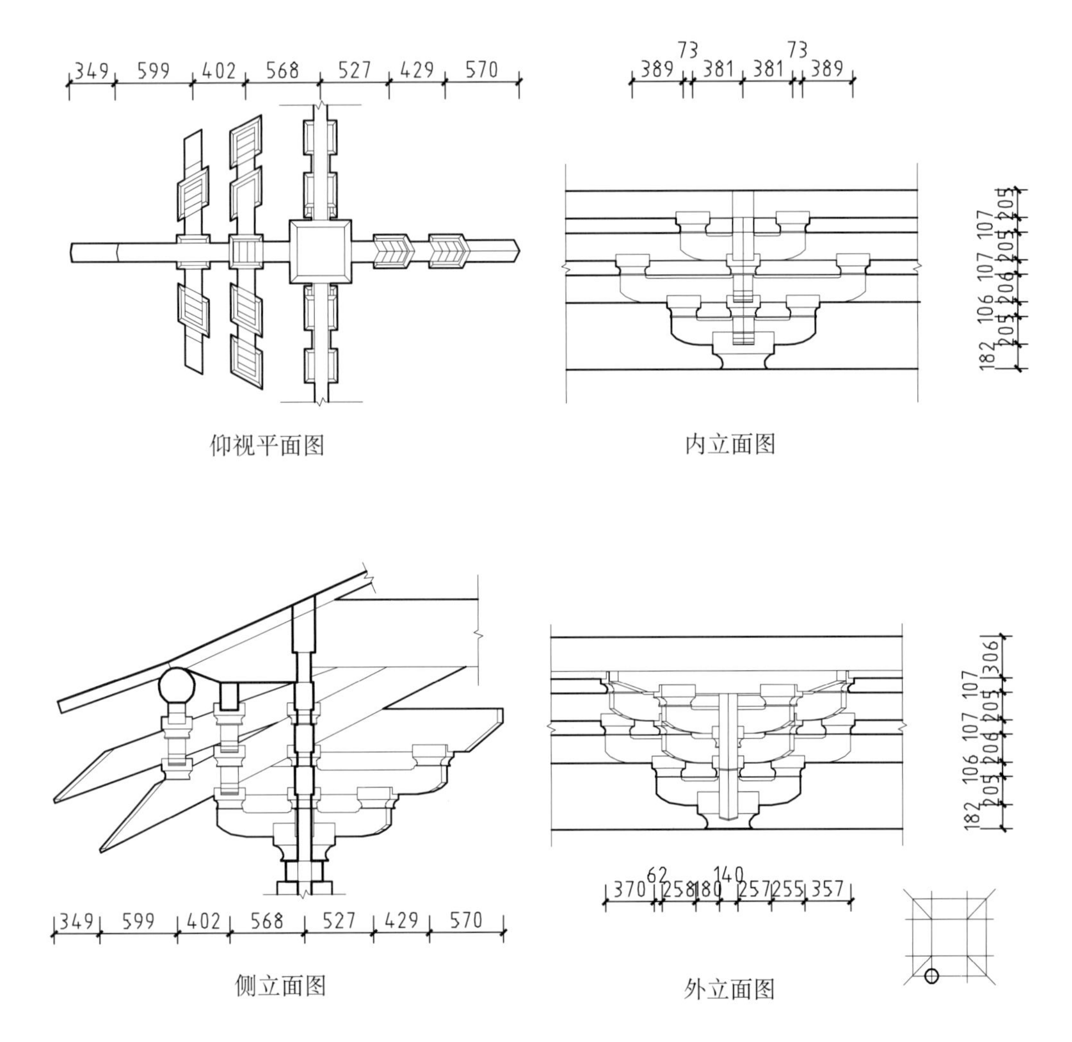

开化寺大殿柱头铺作三视图

Axonometric of Kaihua Monastery's main hall's column-top *puzuo*

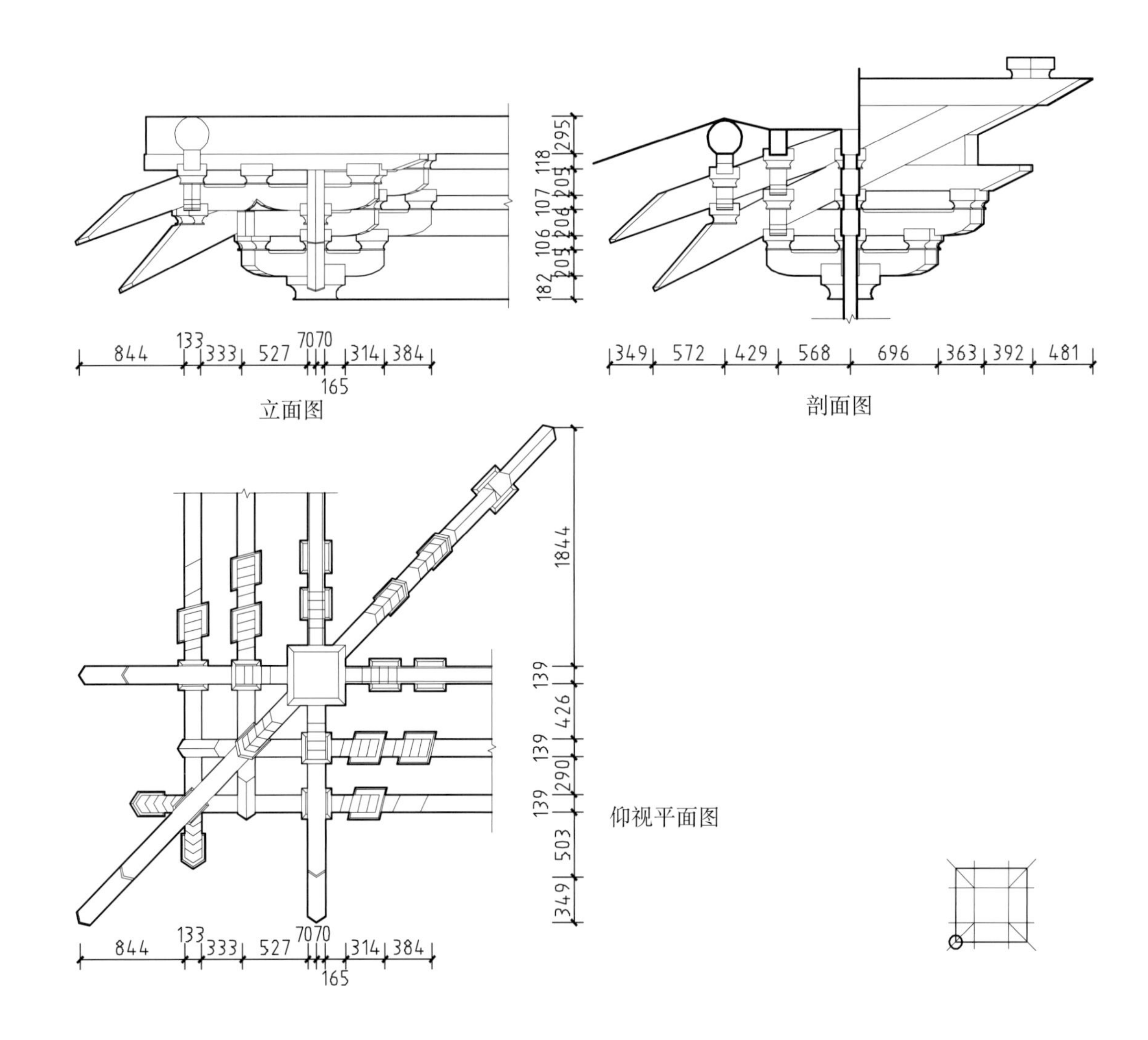

开化寺大殿转角铺作三视图

Axonometric of Kaihua Monastery's main hall's corner *puzuo*

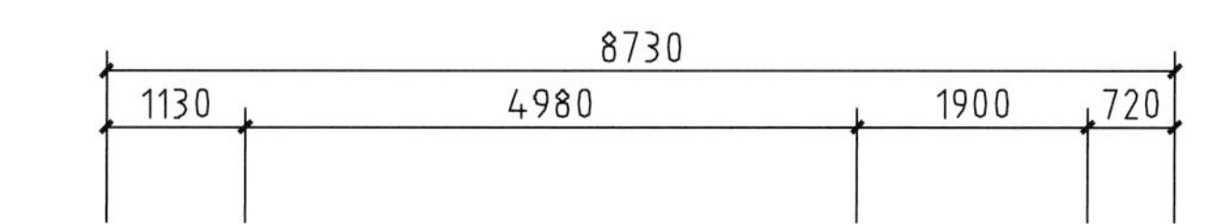

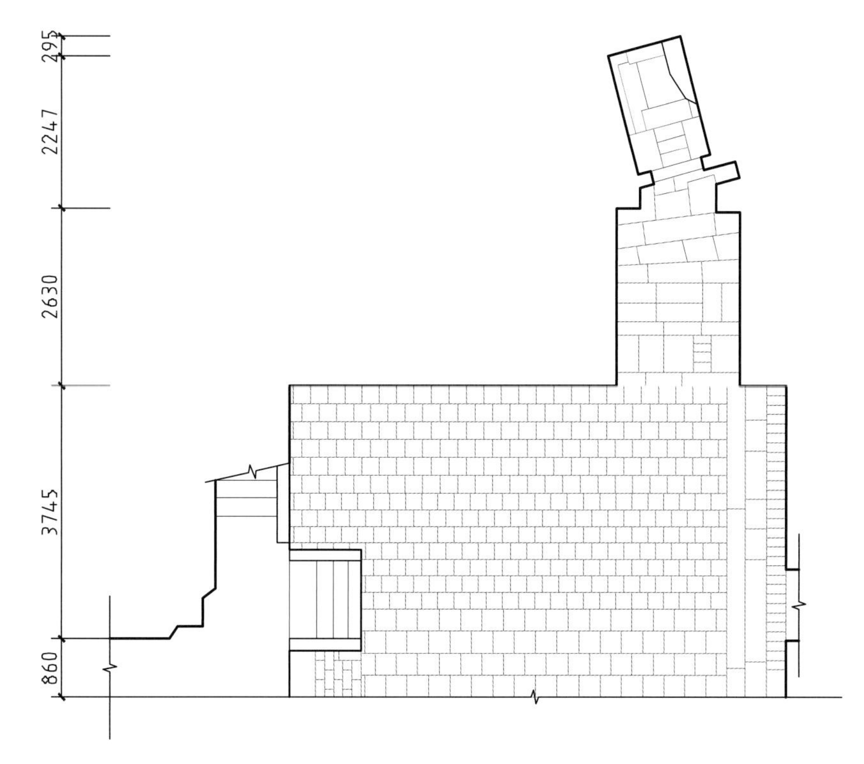

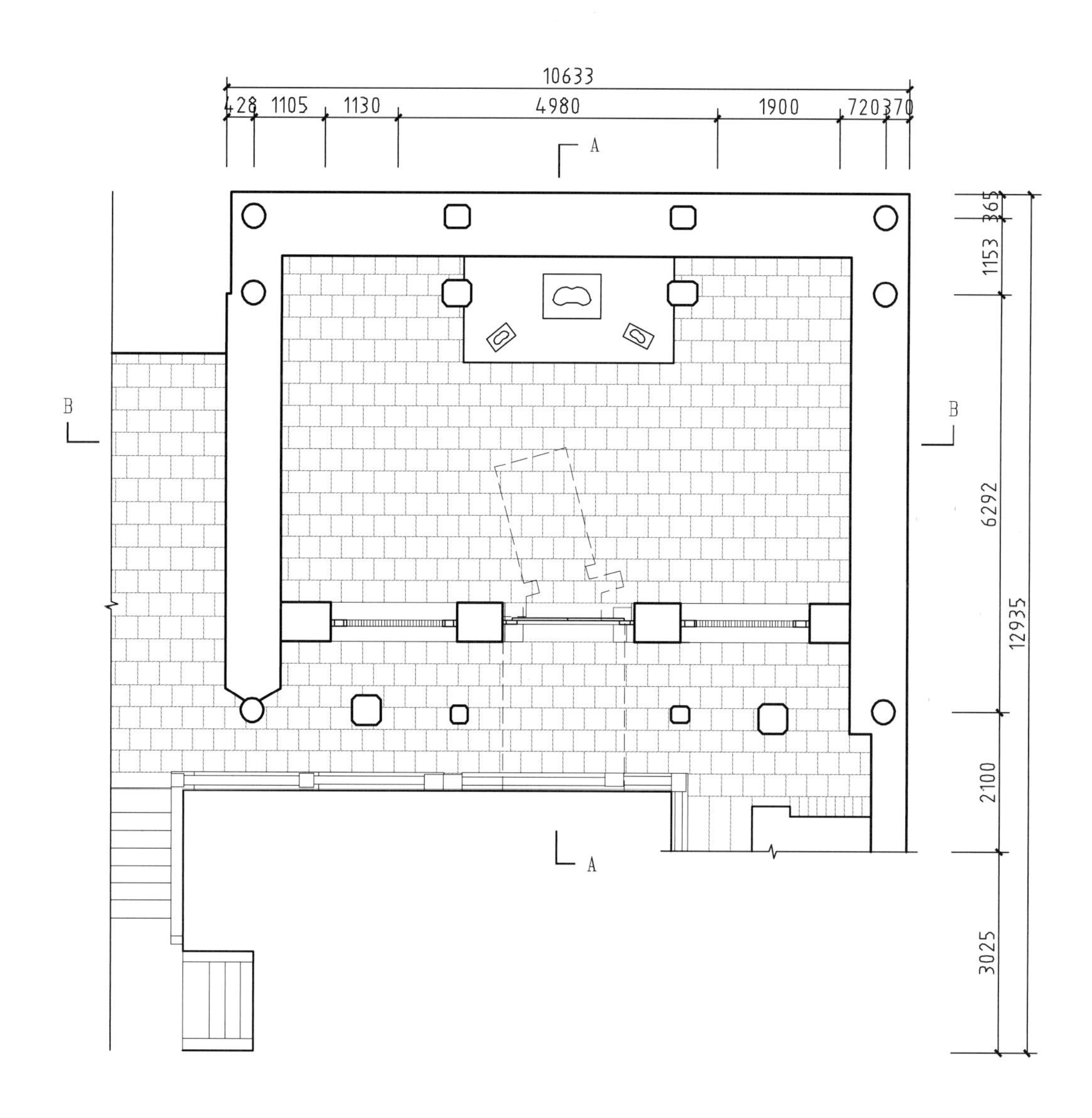

开化寺观音阁一层平面图
Plan of first floor of Kaihua Monastery's Guanyin Pavilion

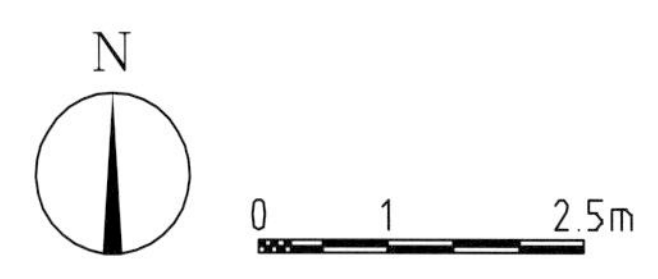

开化寺观音阁二层平面图
Plan of second floor of Kaihua Monastery's Guanyin Pavilion

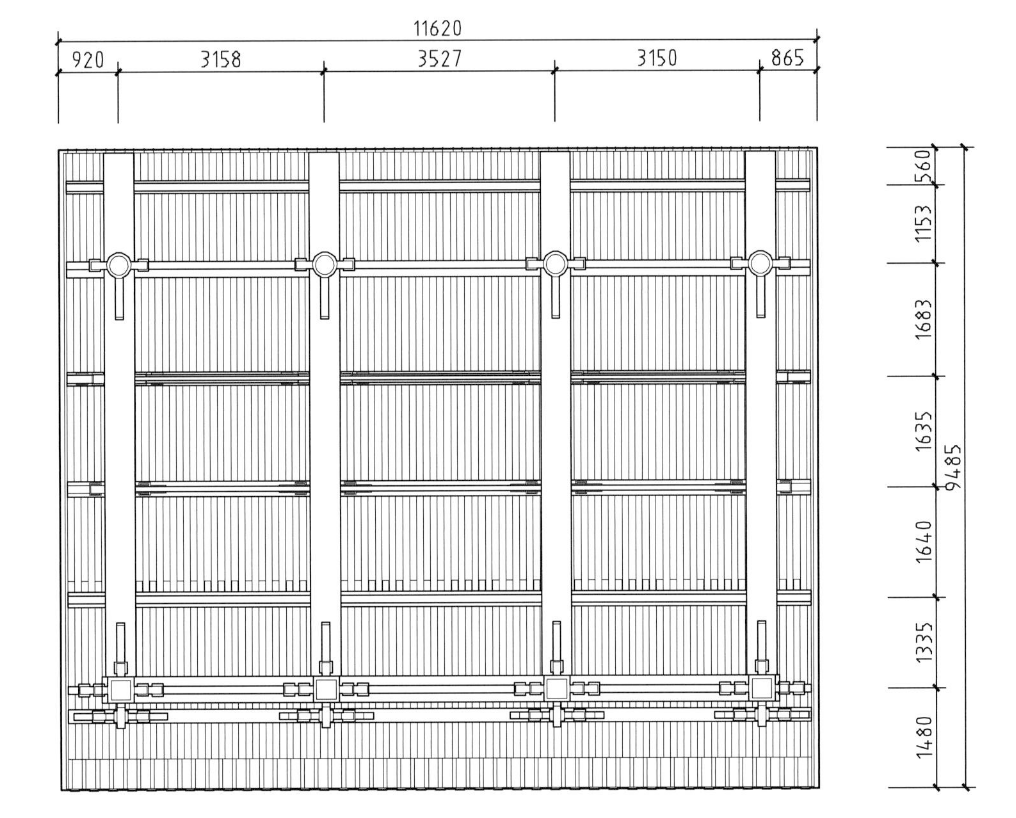

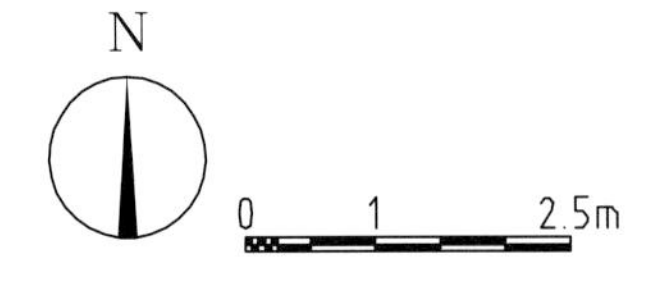

开化寺观音阁梁架仰视平面图

Plan of Kaihua Monastery's Guanyin Pavilion's framework as seen from below

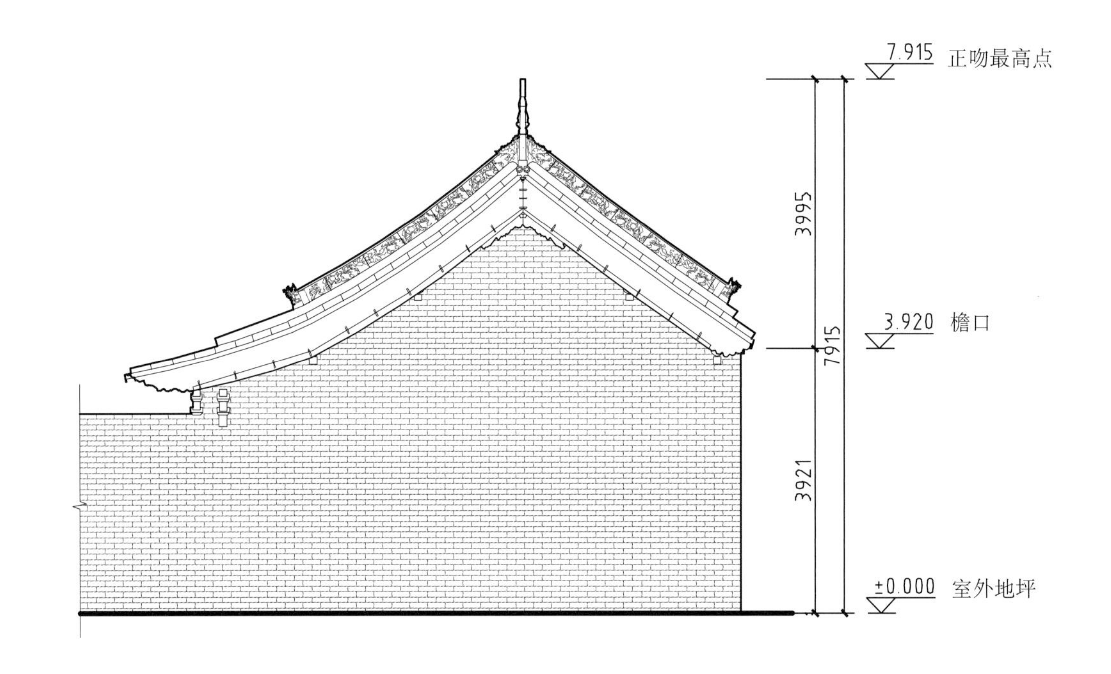

开化寺观音阁东立面图

East elevation of Kaihua Monastery's Guanyin Pavilion

0 1 2.5m

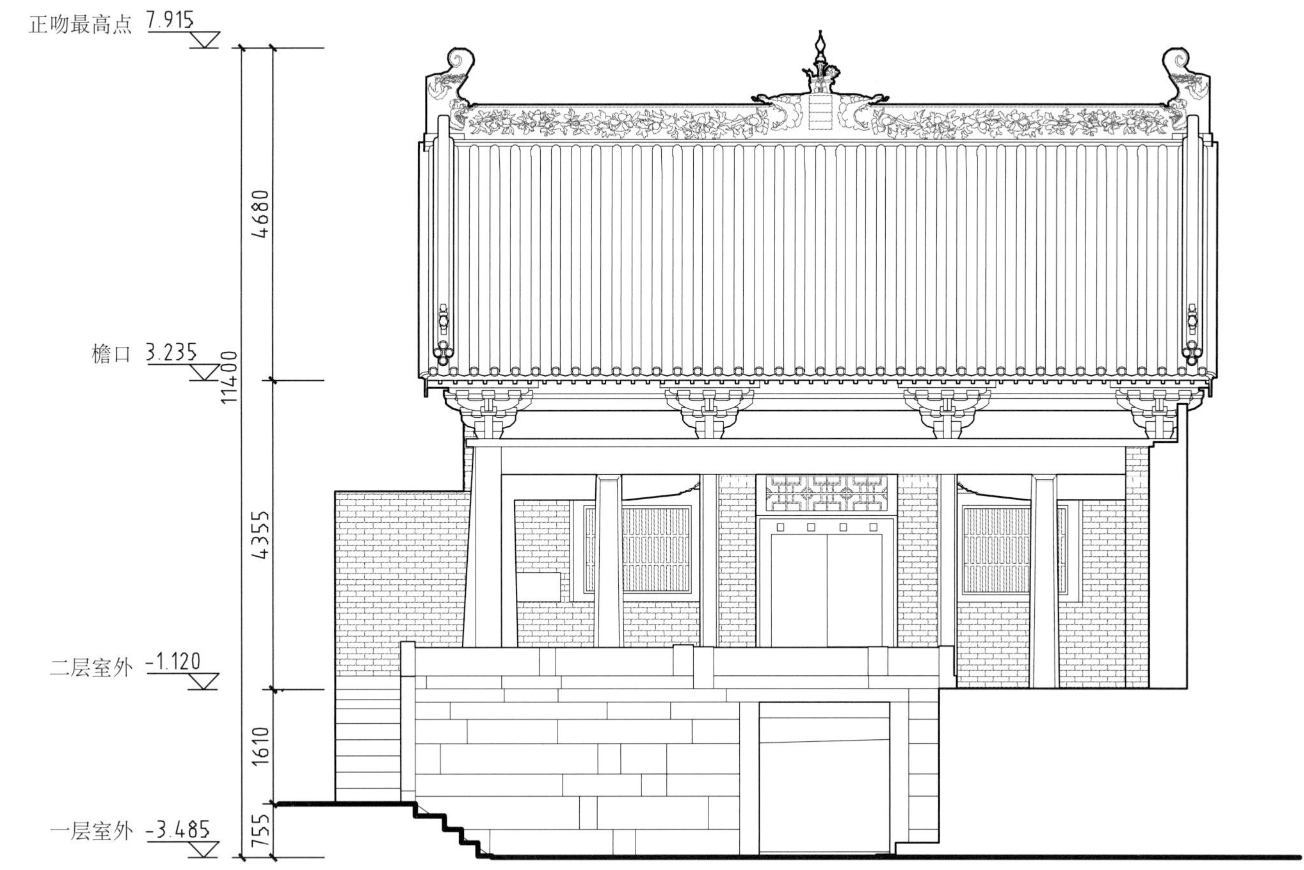

开化寺观音阁南立面图
South elevation of Kaihua Monastery's Guanyin Pavilion

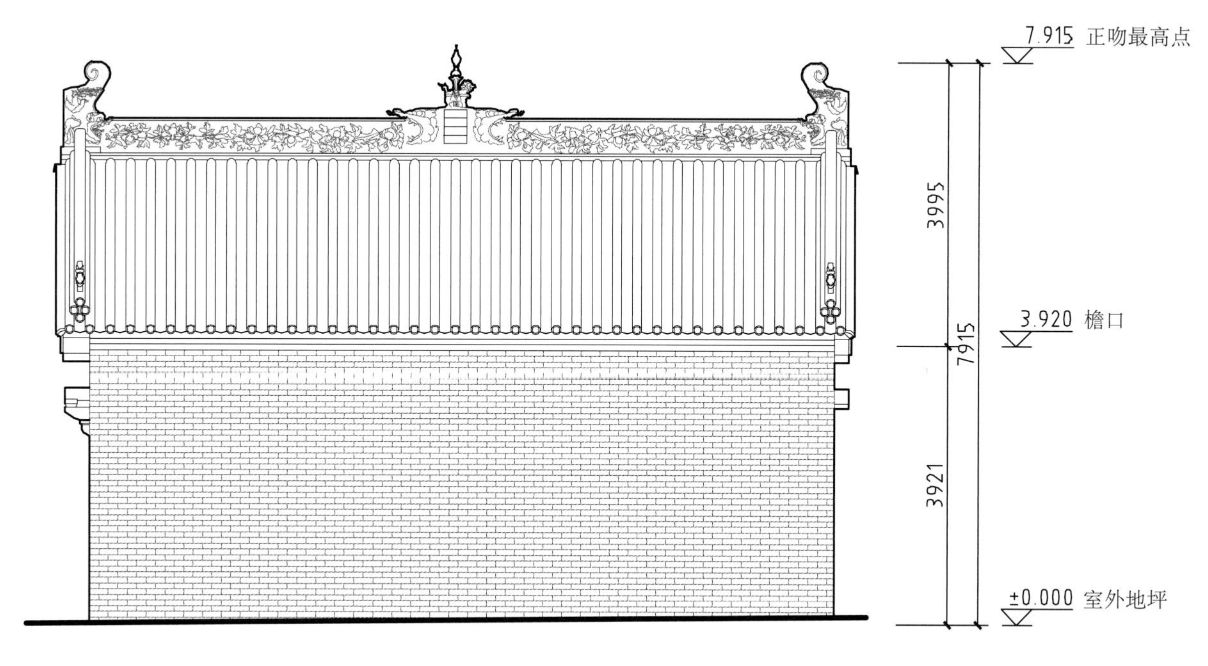

开化寺观音阁北立面图
North elevation of Kaihua Monastery's Guanyin Pavilion

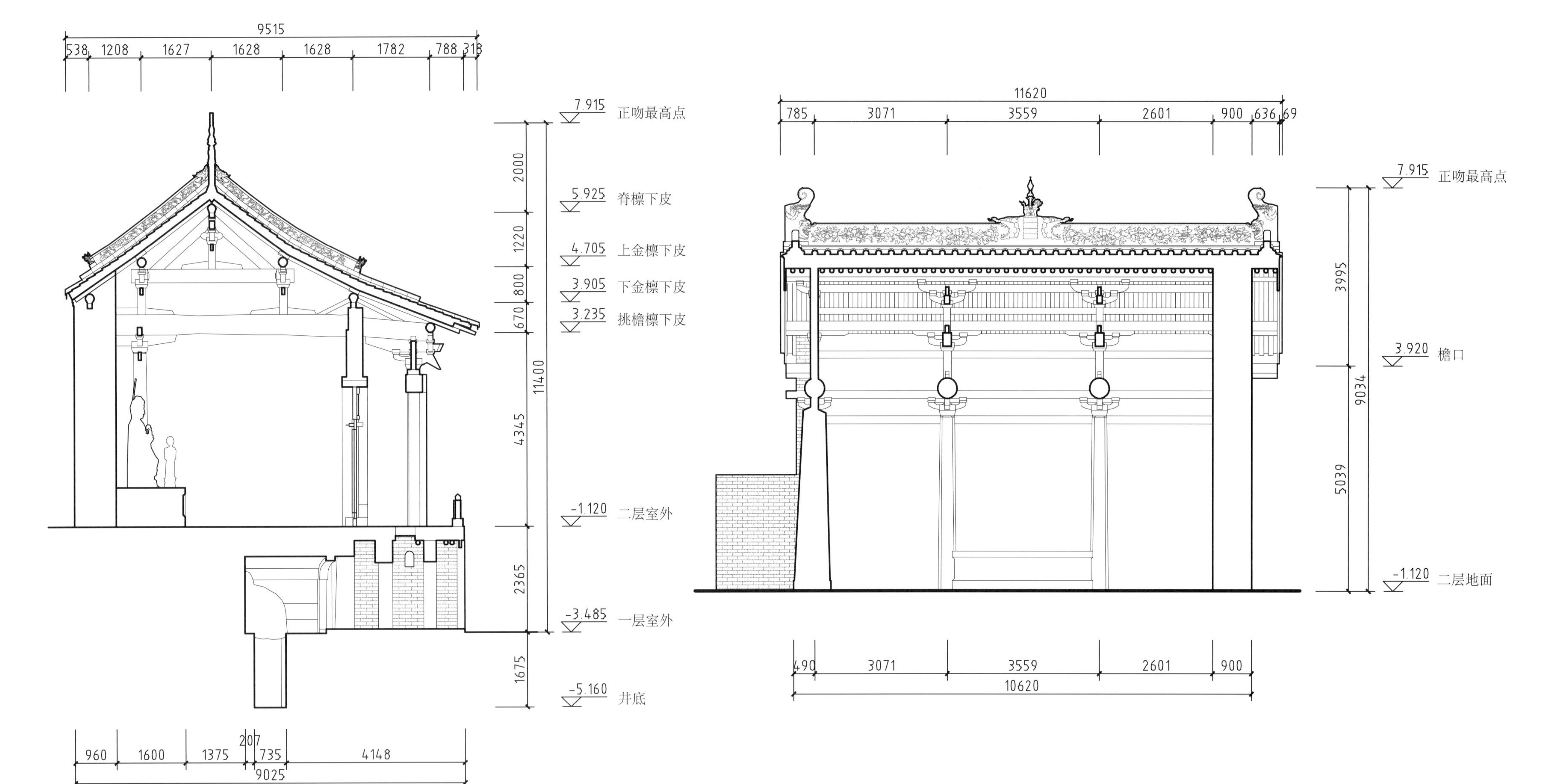

开化寺观音阁 A-A 剖面图
Section A-A of Kaihua Monastery's Guanyin Pavilion

开化寺观音阁 B-B 剖面图
Section B-B of Kaihua Monastery's Guanyin Pavilion

资圣寺
Zisheng Monastery

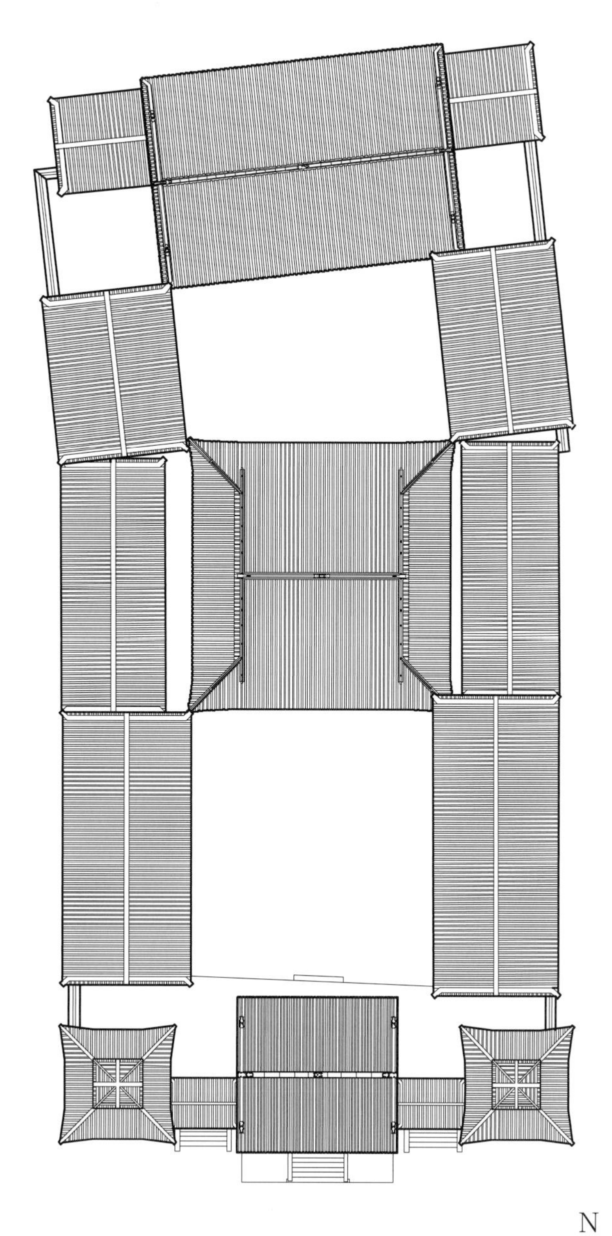

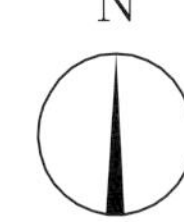

资圣寺屋顶总平面图
Site roof plan of Zisheng Monastery

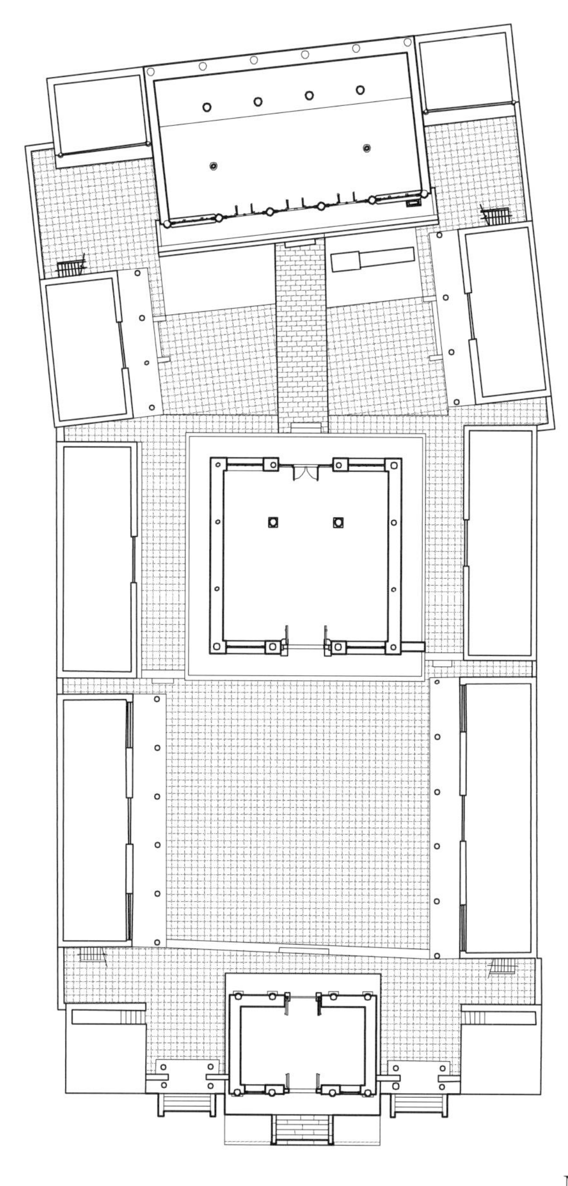

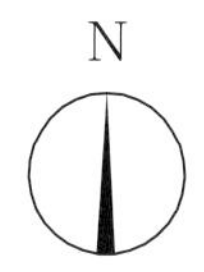

资圣寺总平面图
Site plan of Zisheng Monastery

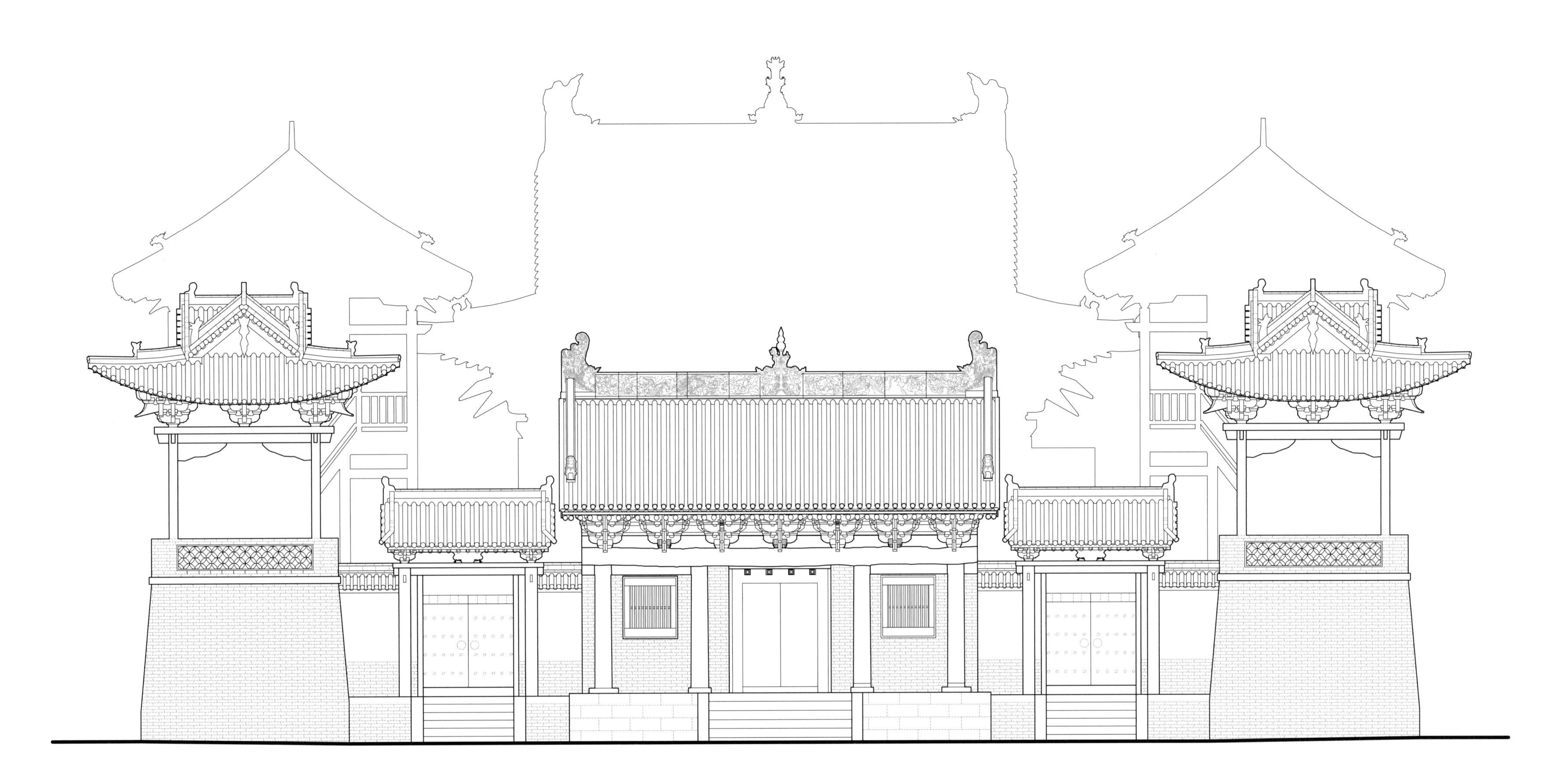

资圣寺南侧总平面图
South site elevation of Zisheng Monastery

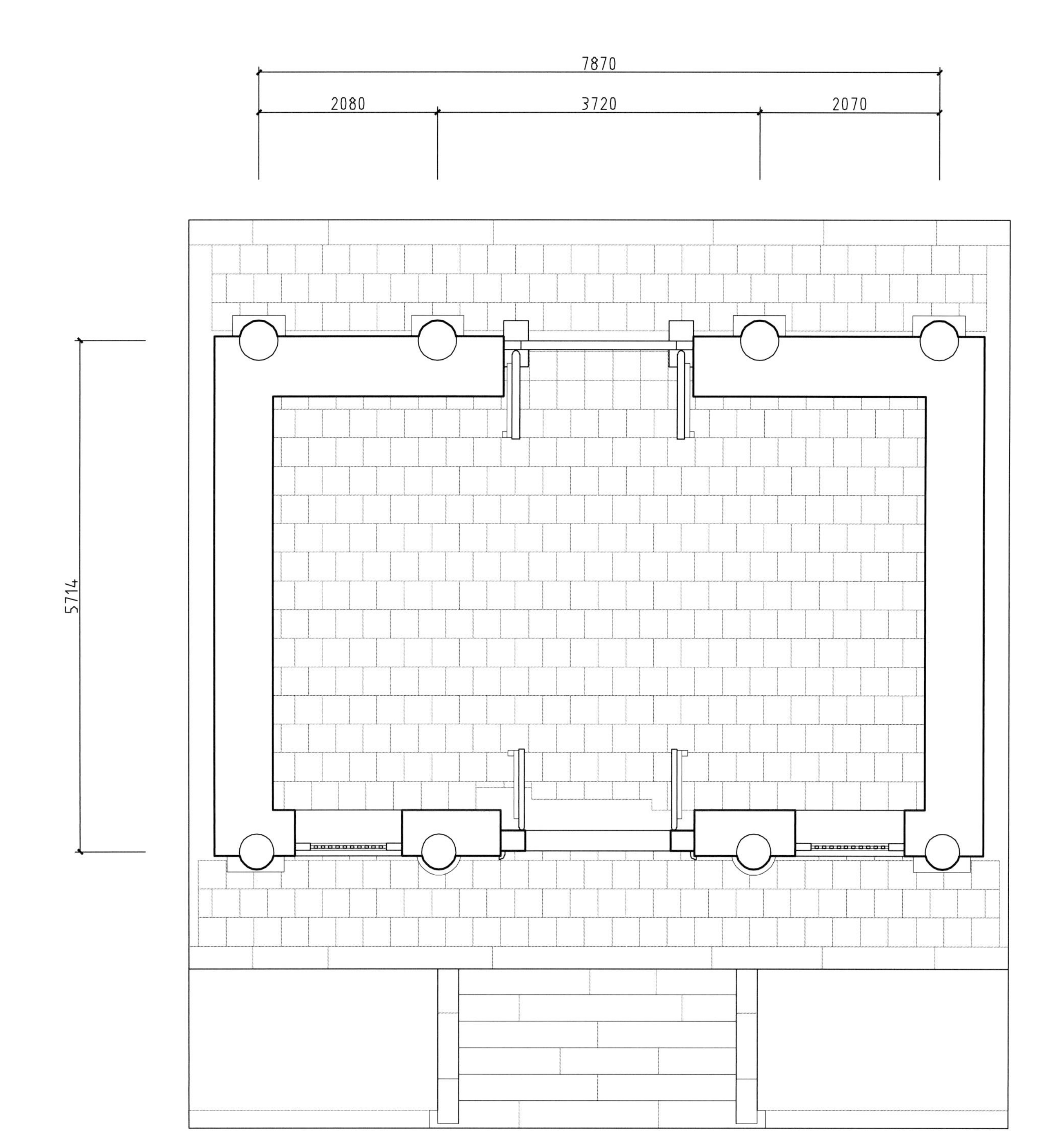

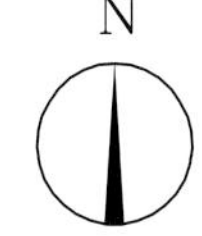

资圣寺山门平面图

Plan of Zisheng Monastery's *shanmen*

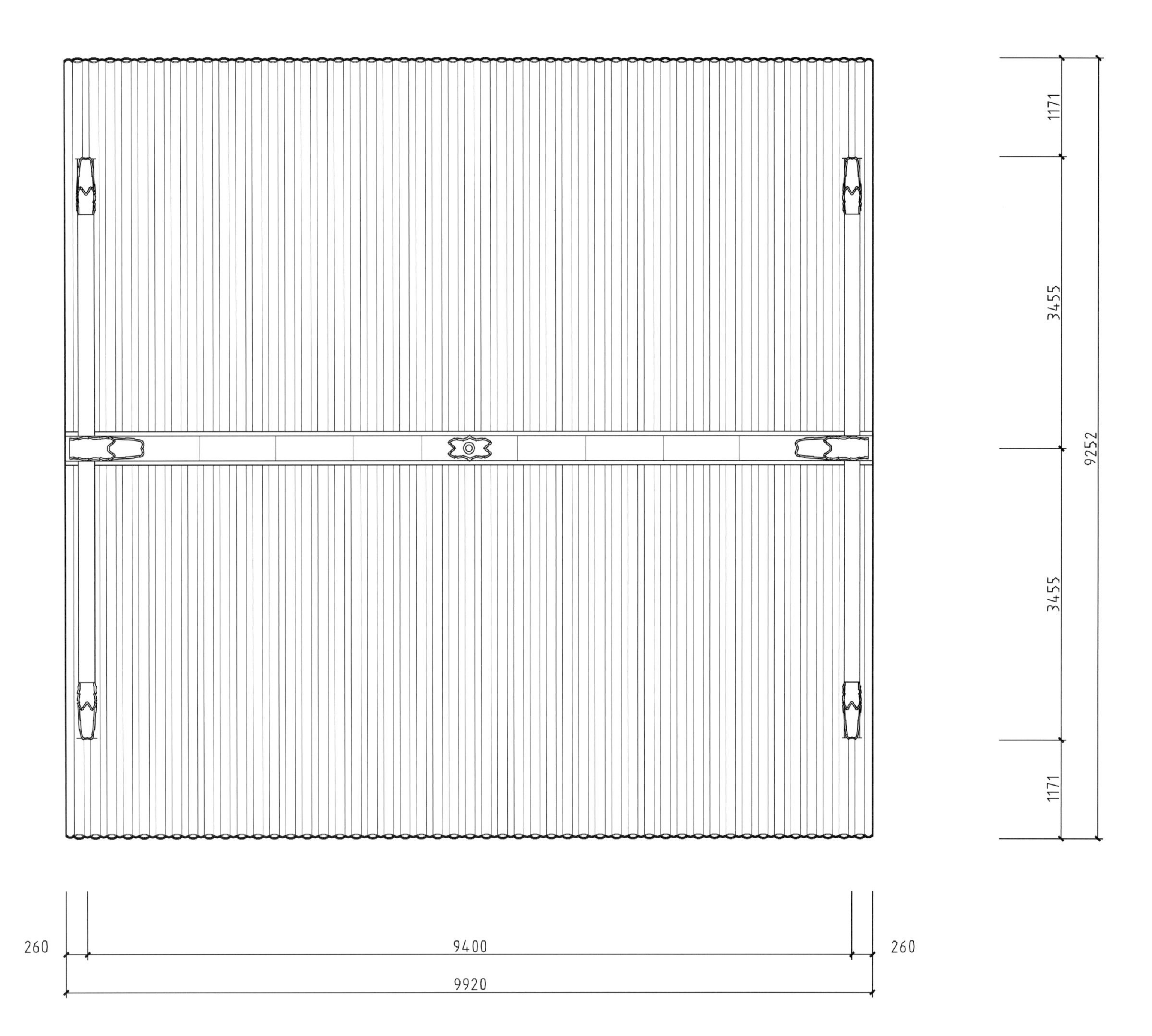

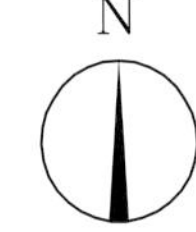

资圣寺山门屋顶平面图

Roof plan of Zisheng Monastery's *shanmen*

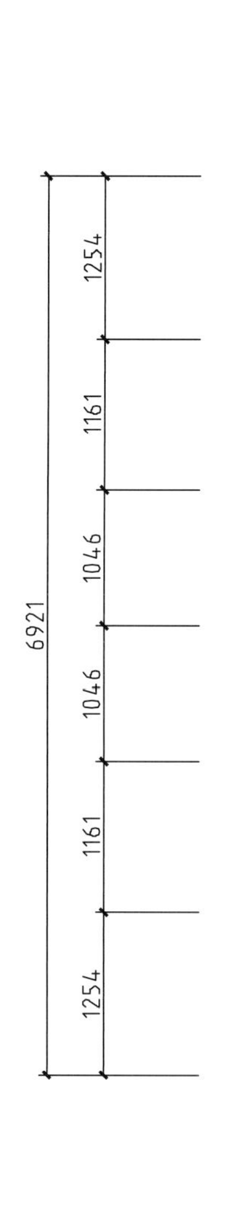

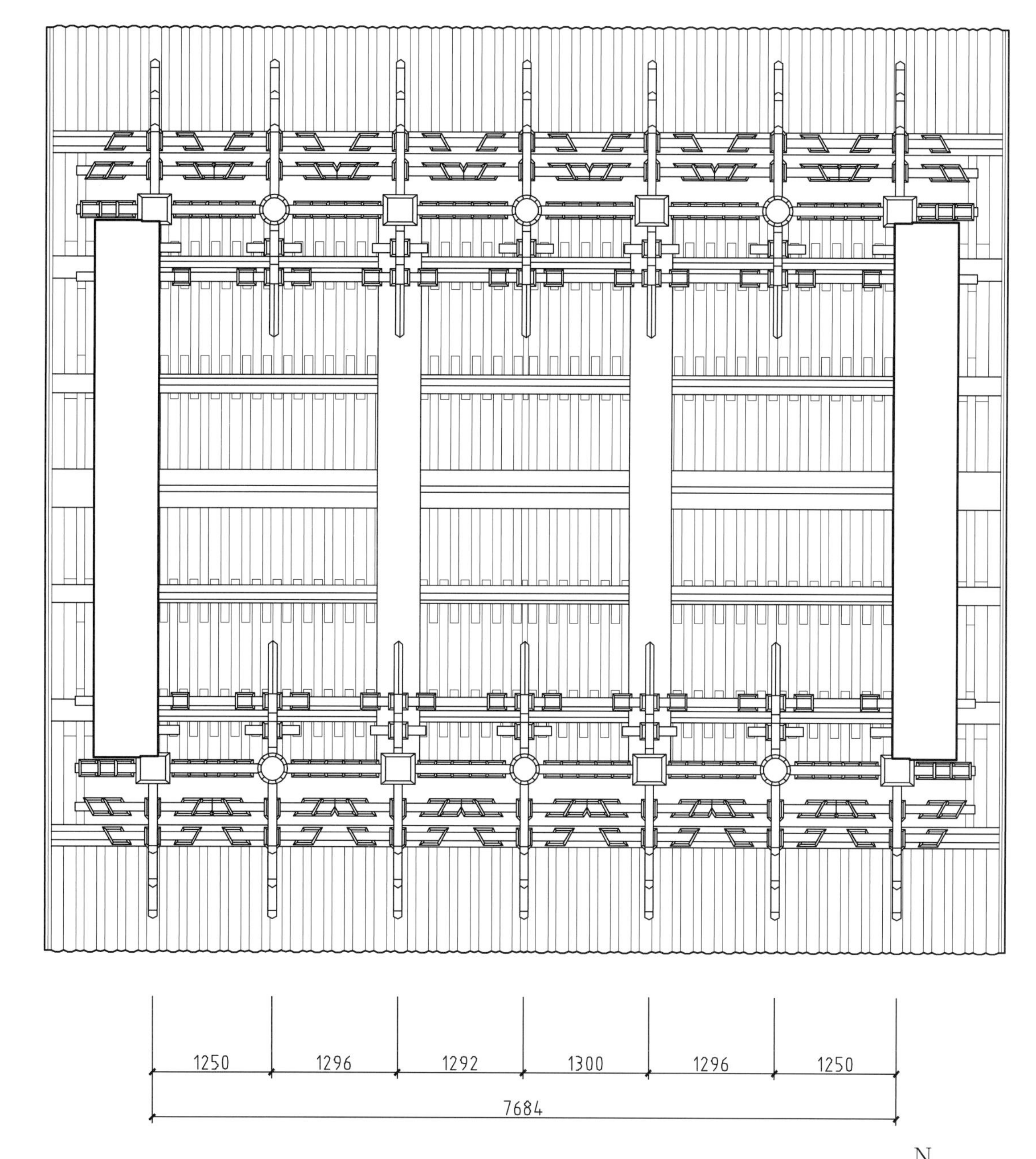

资圣寺山门梁架仰视平面图

Plan of Zisheng Monastery's shanmen's framework as seen from below

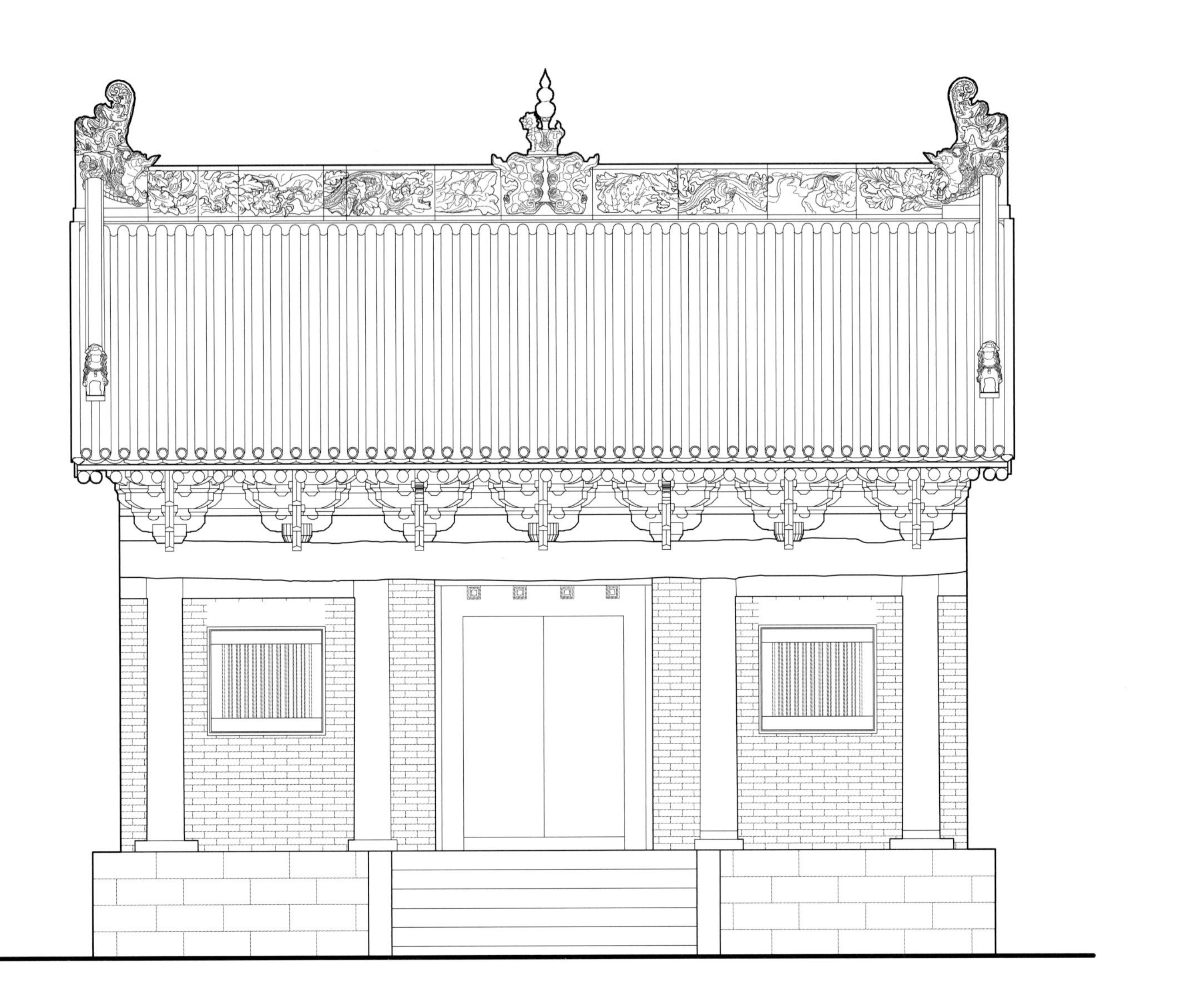

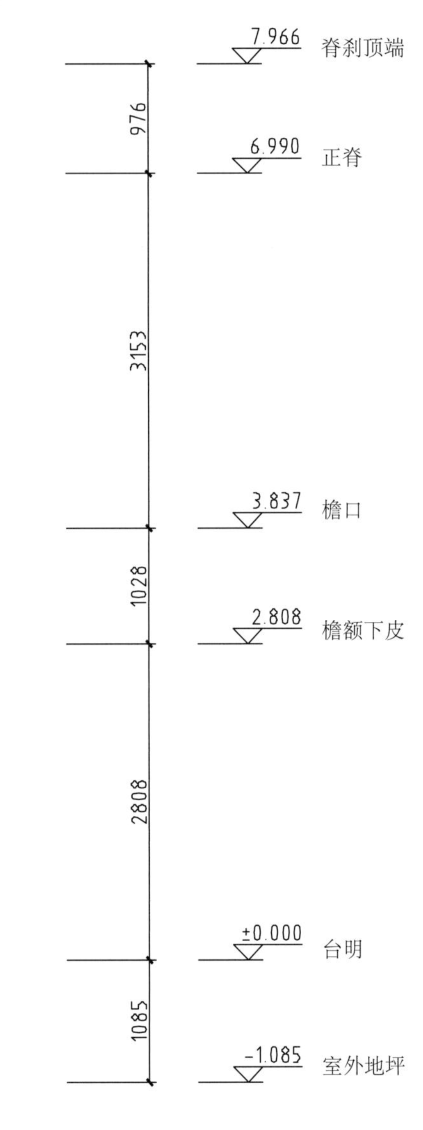

资圣寺山门正立面图

Front elevation of Zisheng Monastery's *shanmen*

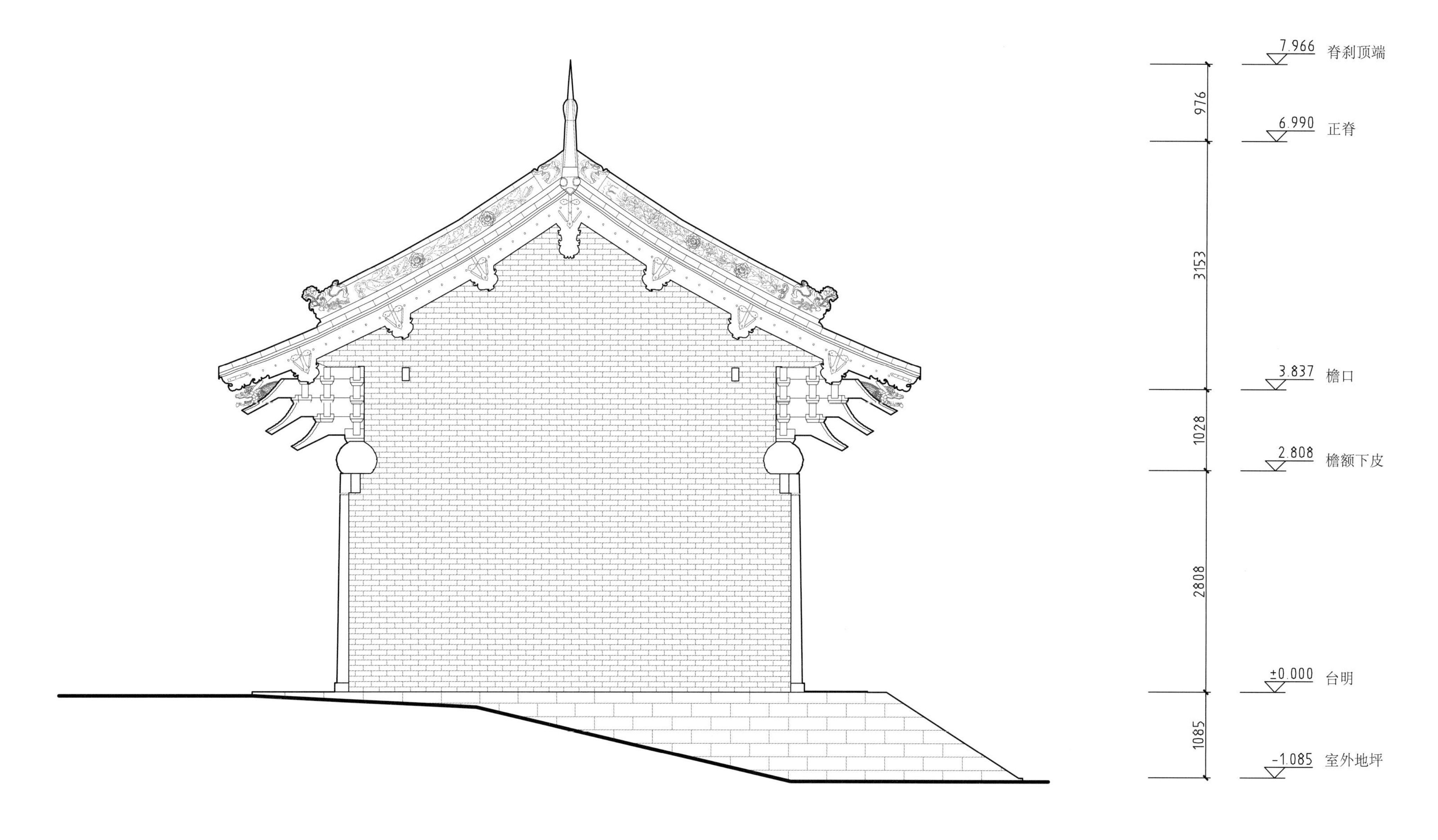

资圣寺山门侧立面图

Side elevation of Zisheng Monastery's *shanmen*

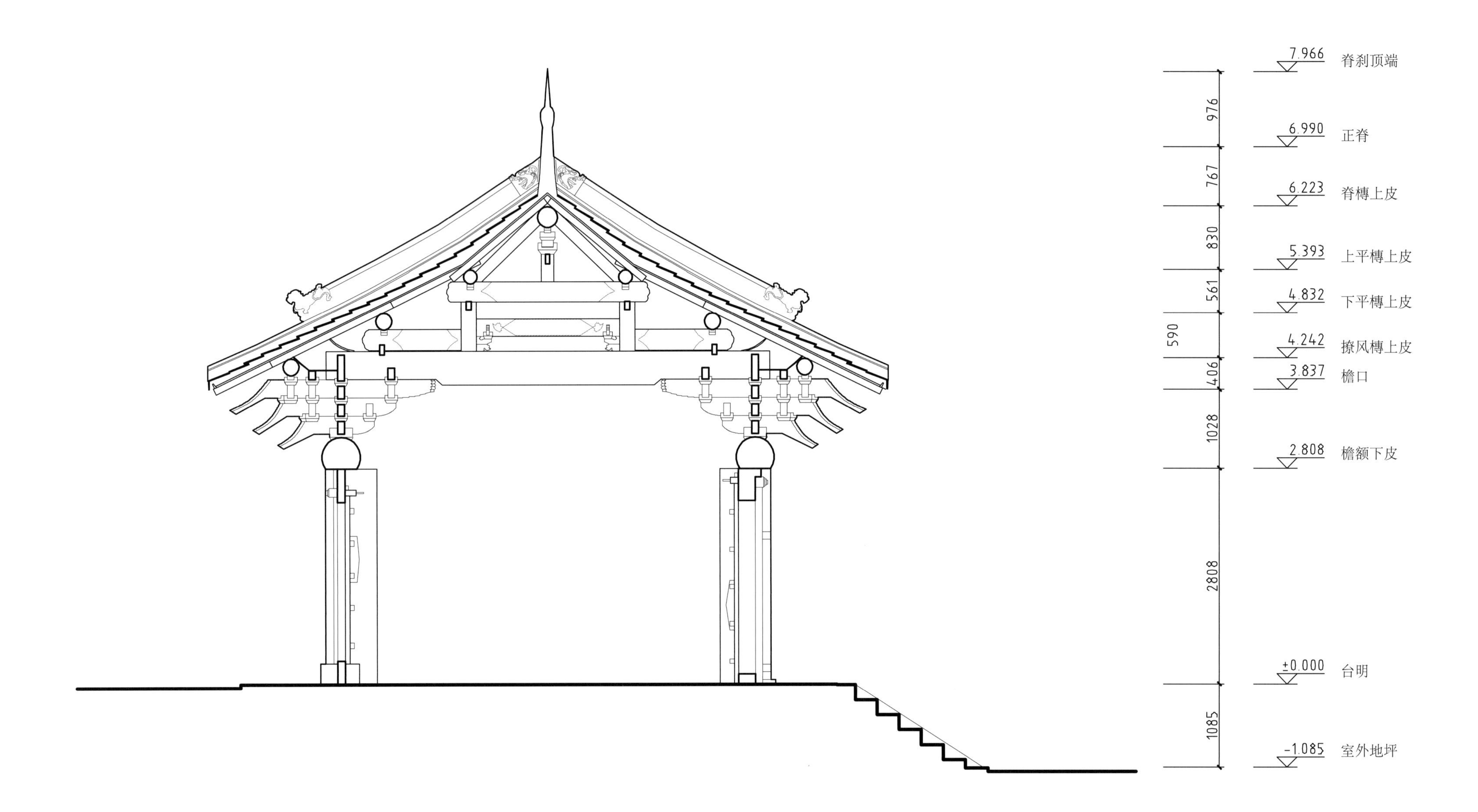

资圣寺山门横剖面图

Cross-section of Zisheng Monastery's *shanmen*

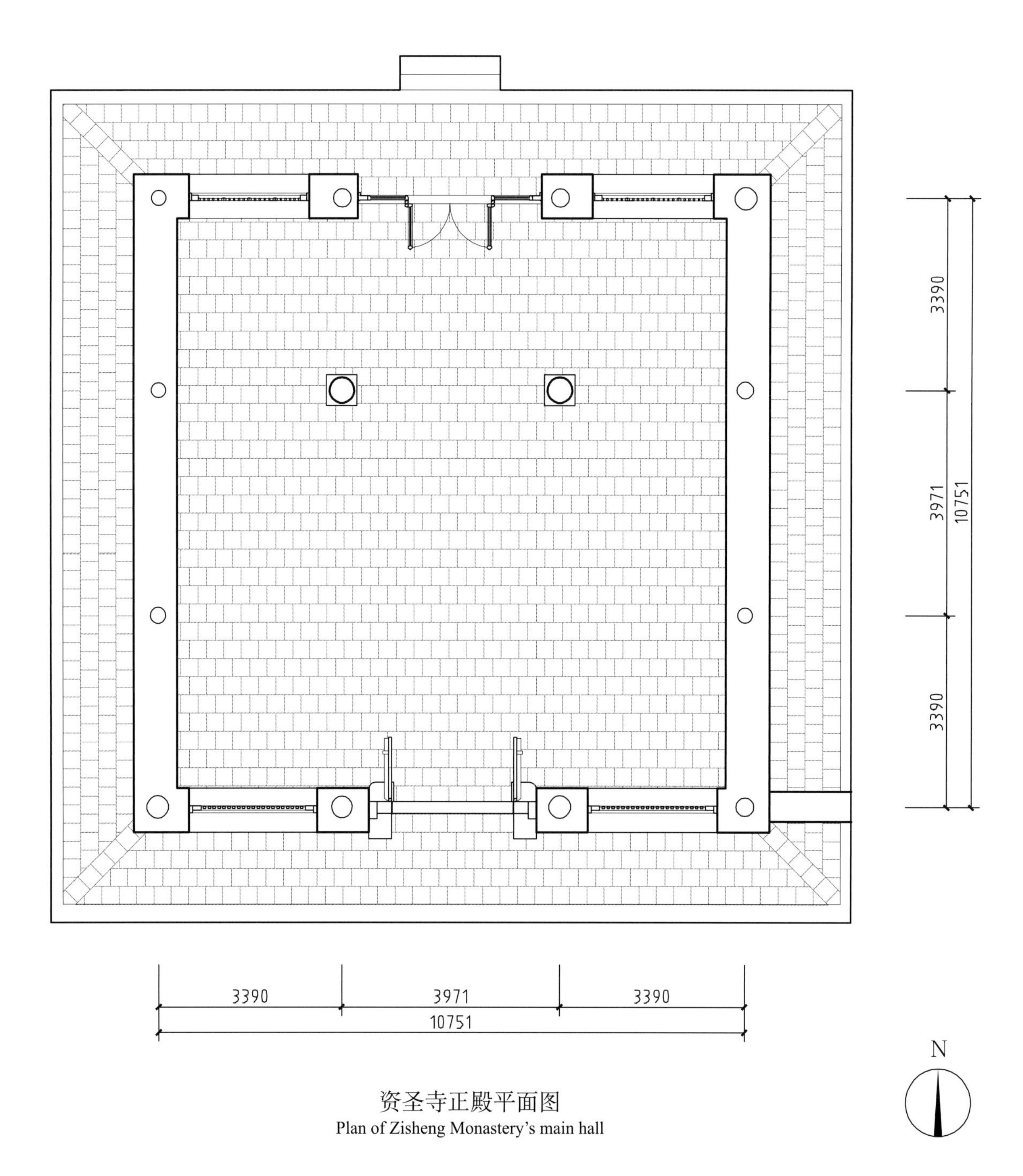

资圣寺正殿平面图
Plan of Zisheng Monastery's main hall

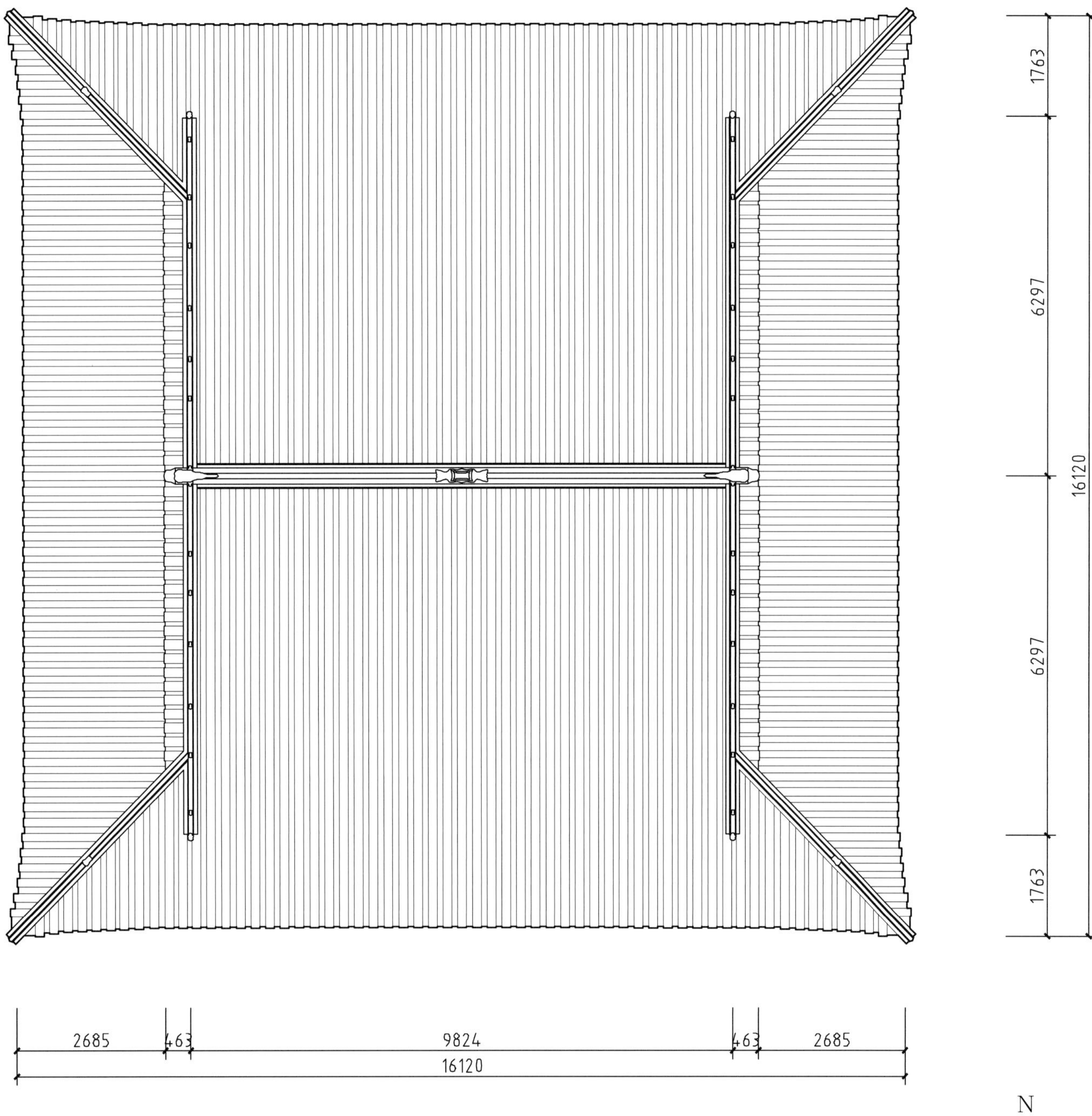

资圣寺正殿屋顶平面图

Roof plan of Zisheng Monastery's main hall

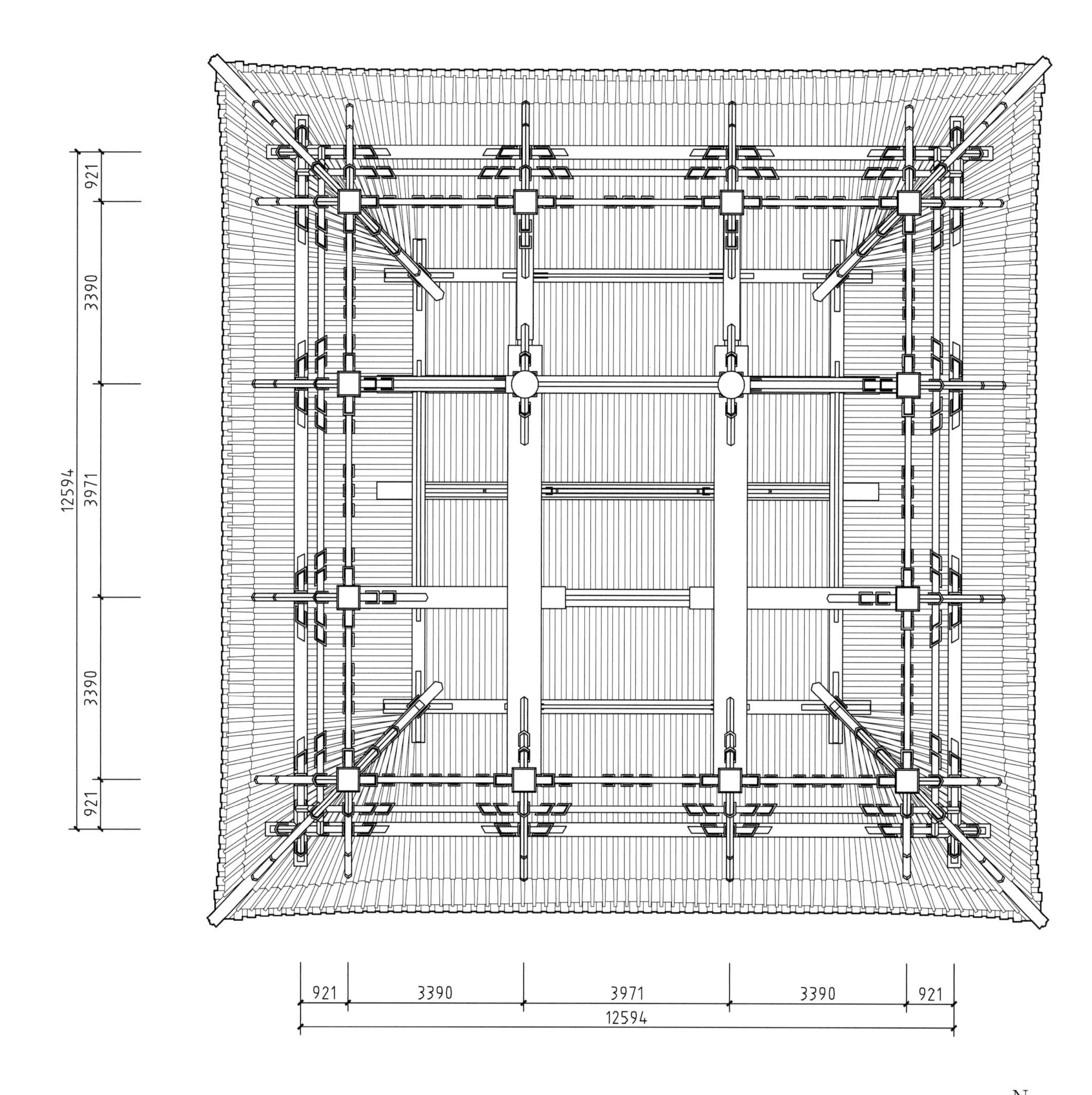

资圣寺正殿梁架仰视平面图

Plan of Zisheng Monastery's main hall's framework as seen from below

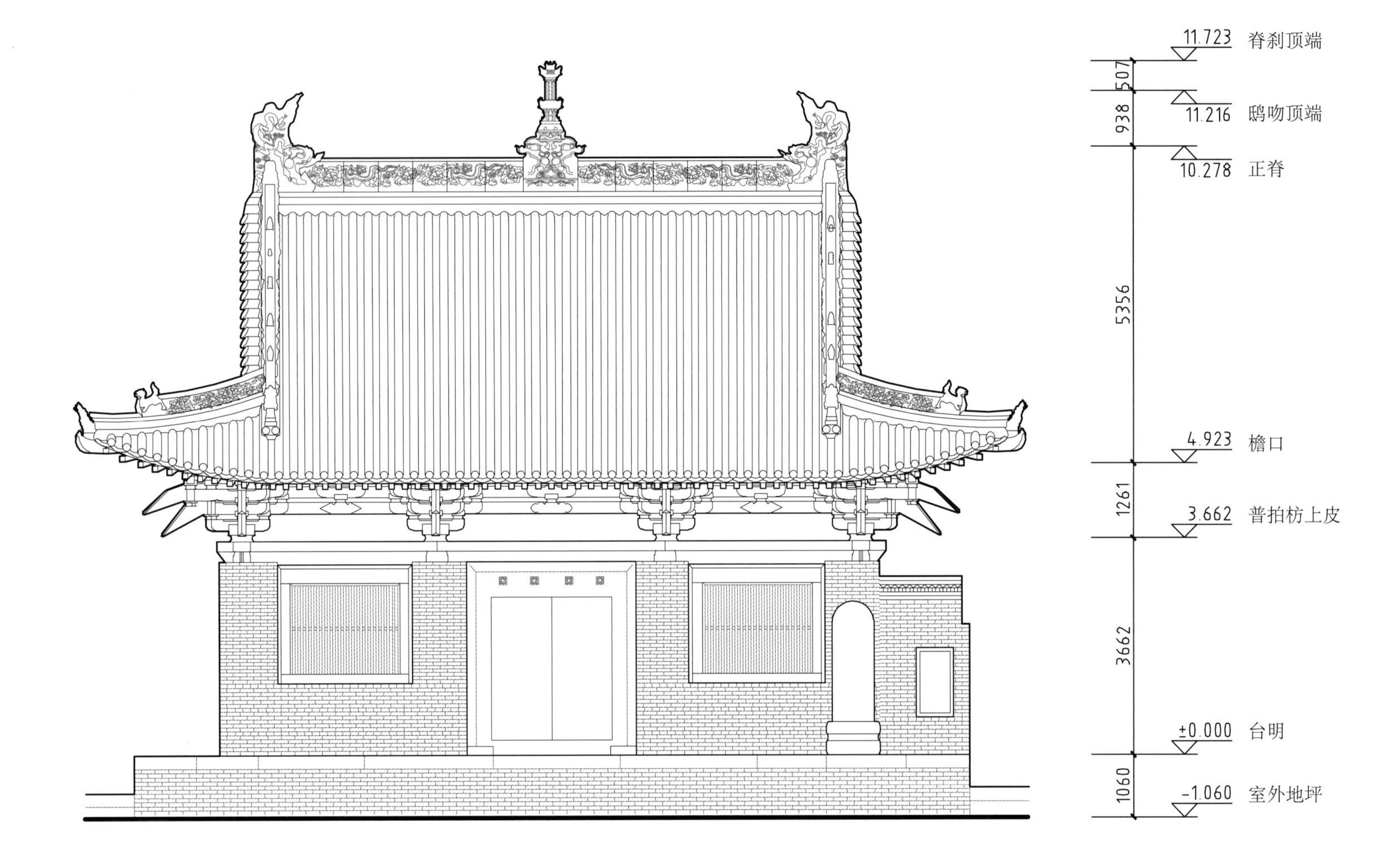

资圣寺正殿正立面图
Front elevation of Zisheng Monastery's main hall

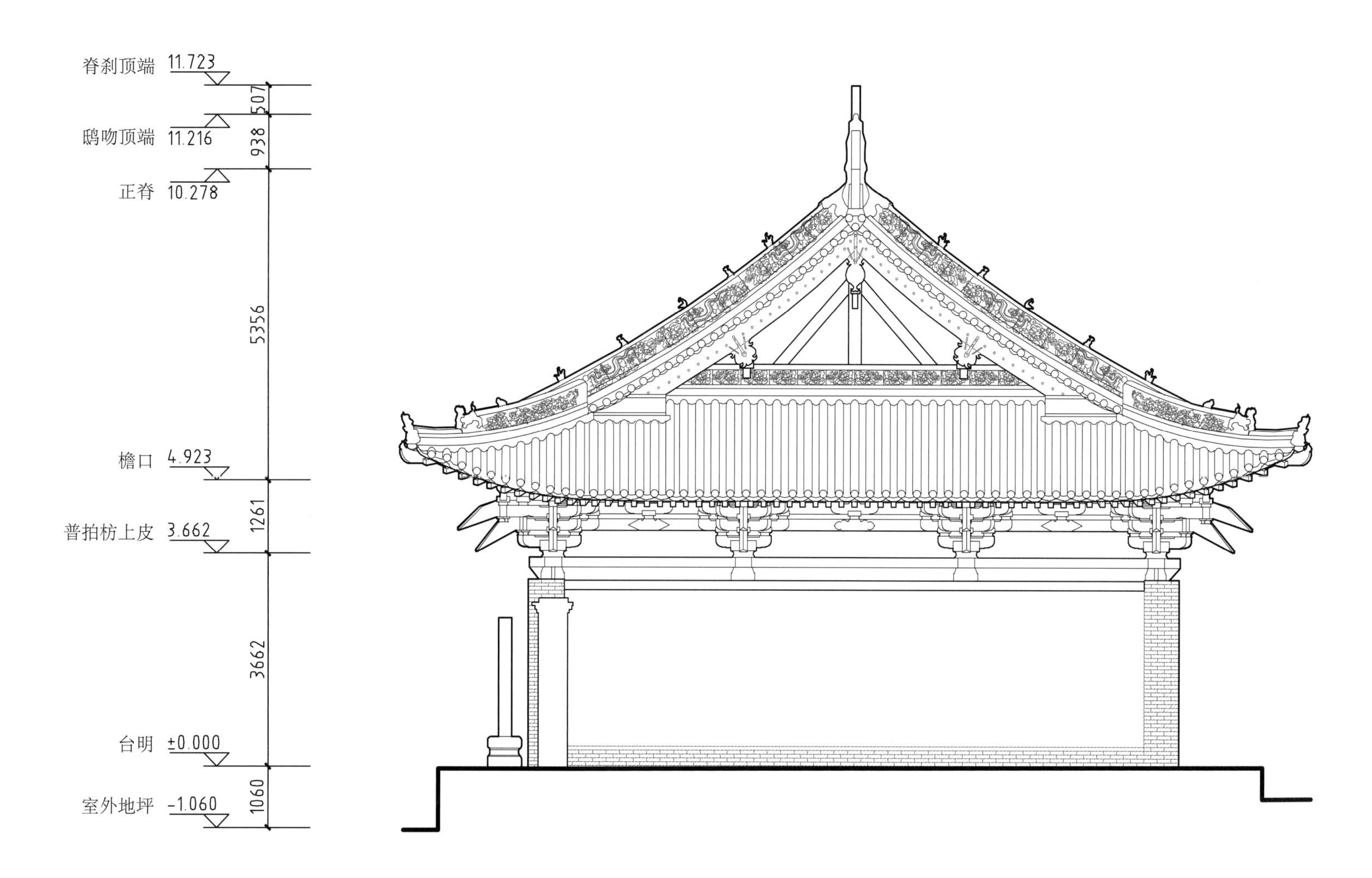

资圣寺正殿侧立面图

Side elevation of Zisheng Monastery's main hall

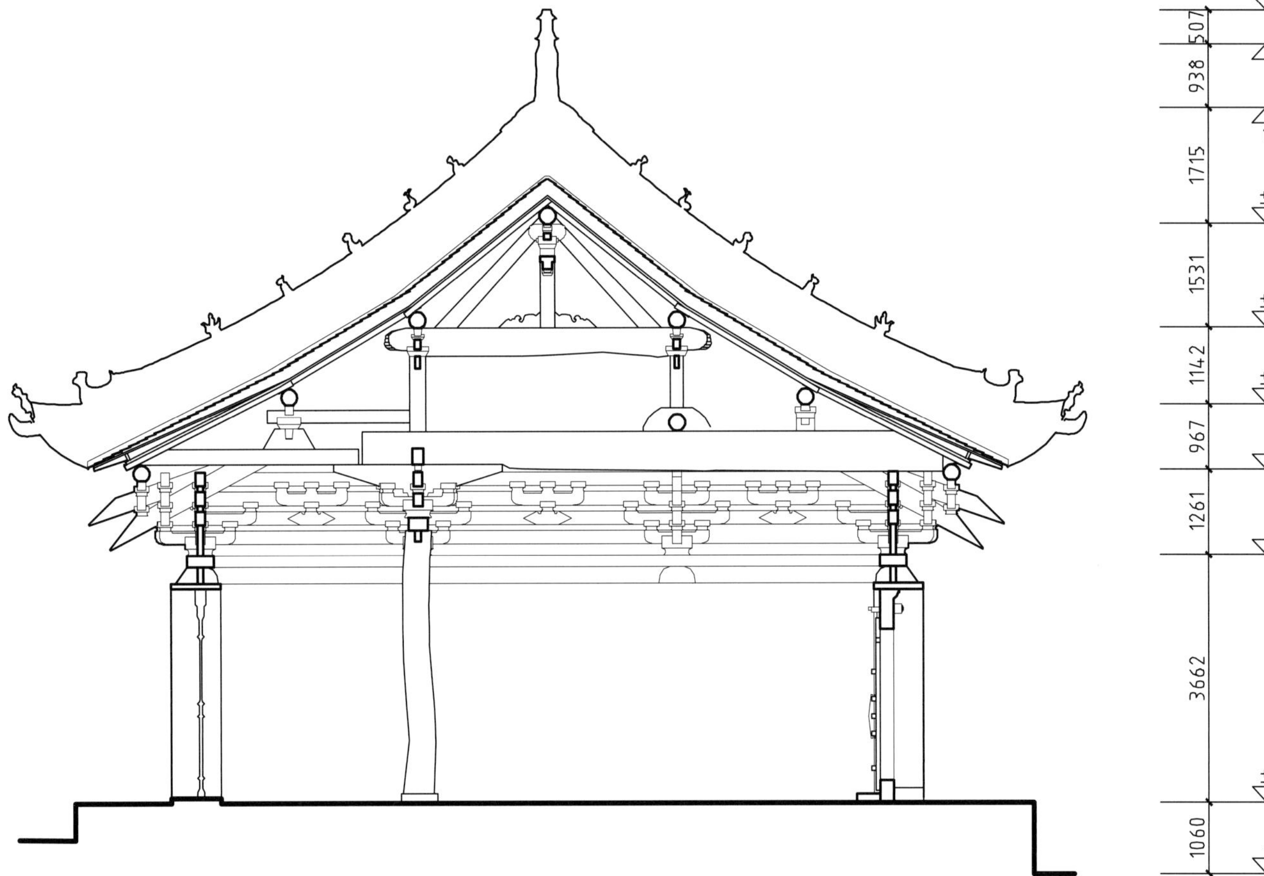

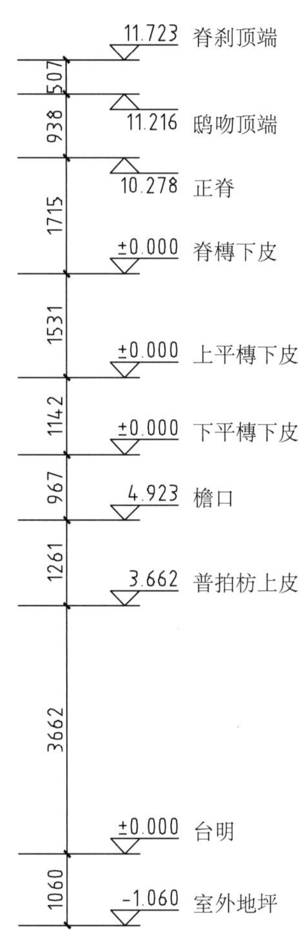

资圣寺正殿横剖面图
Cross-section of Zisheng Monastery's main hall

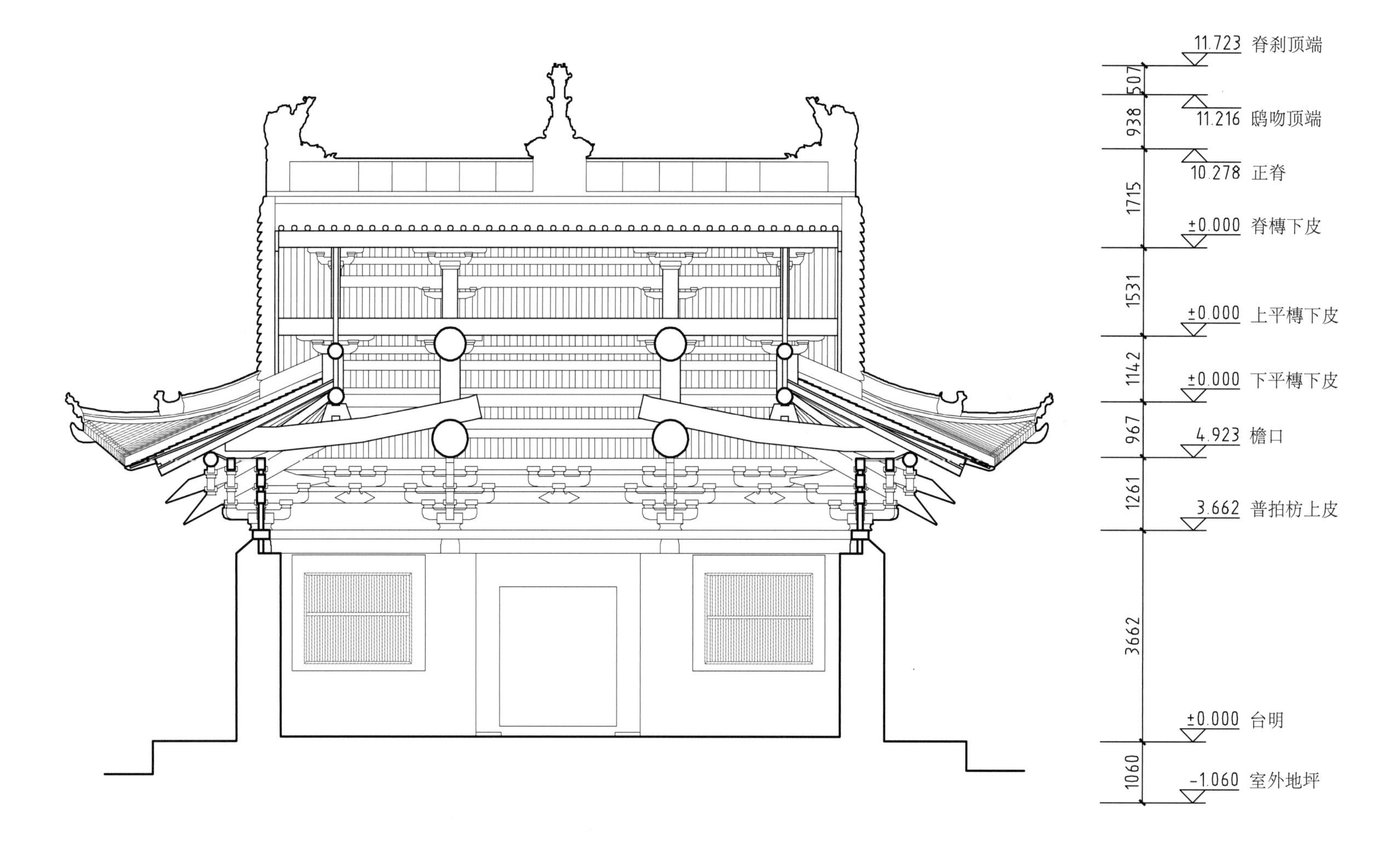

资圣寺正殿纵剖面图

Longitudinal section of Zisheng Monastery's main hall

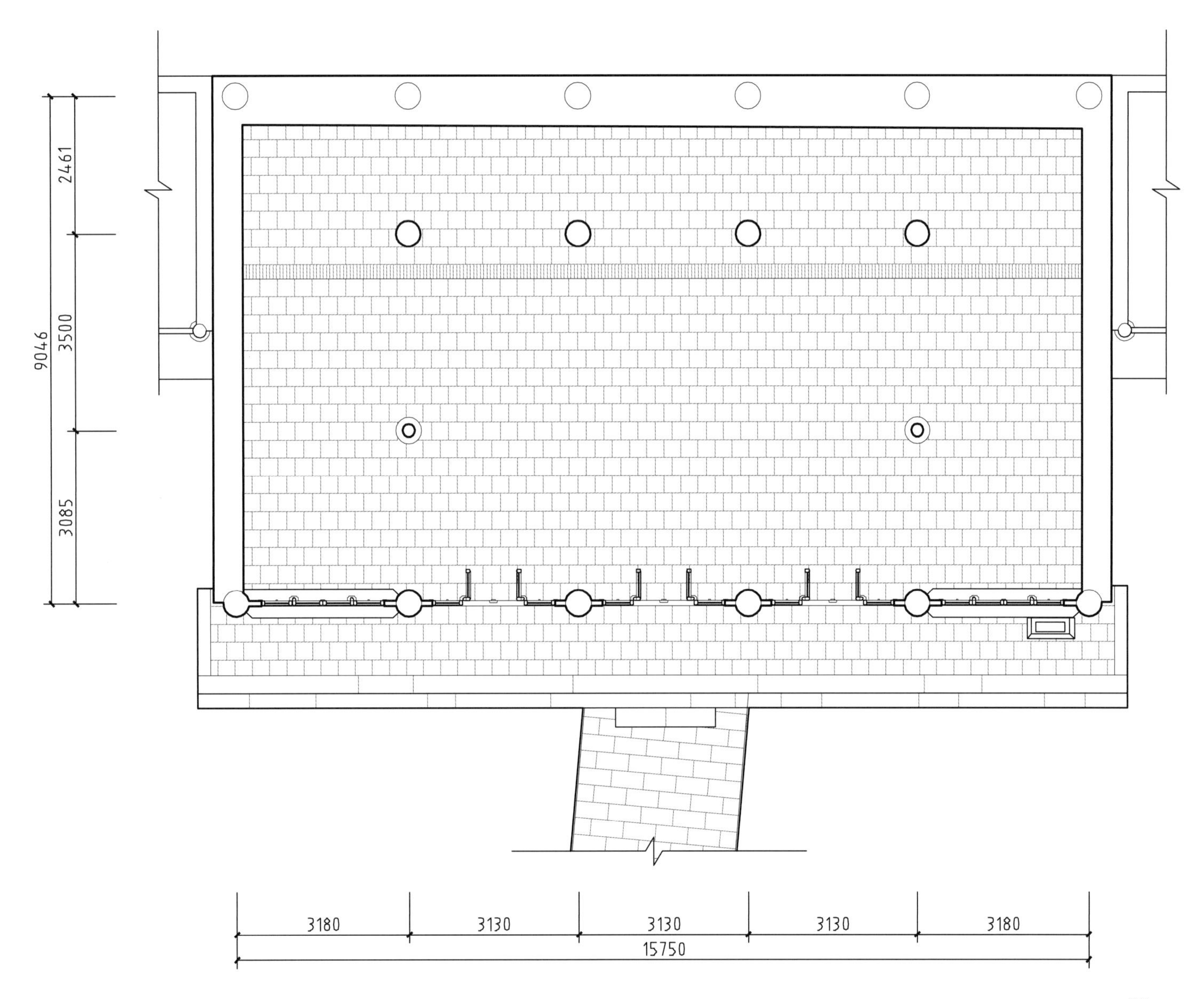

资圣寺后殿平面图

Plan of Zisheng Monastery's rear hall

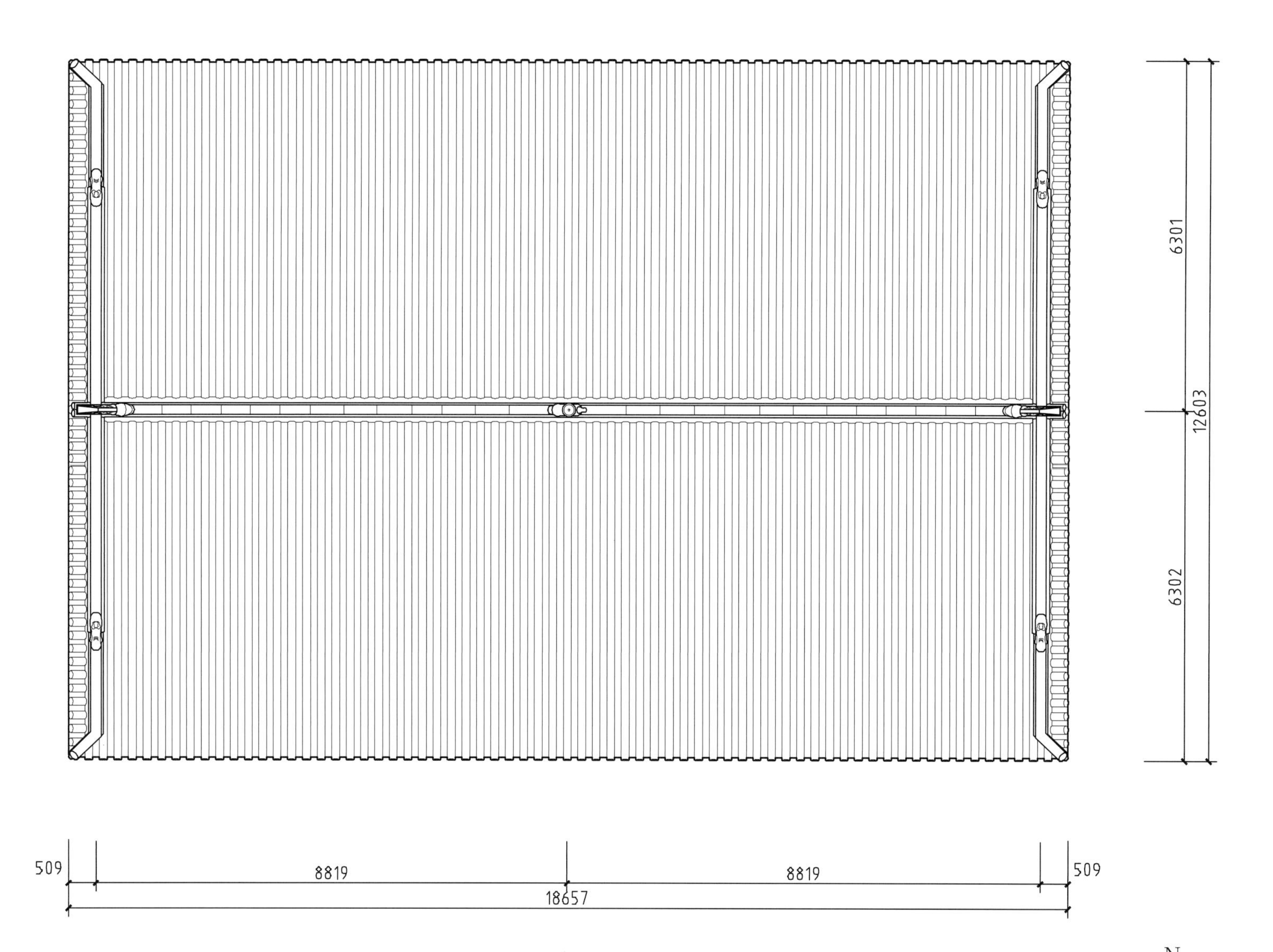

资圣寺后殿屋顶平面图
Roof plan of Zisheng Monastery's rear hall

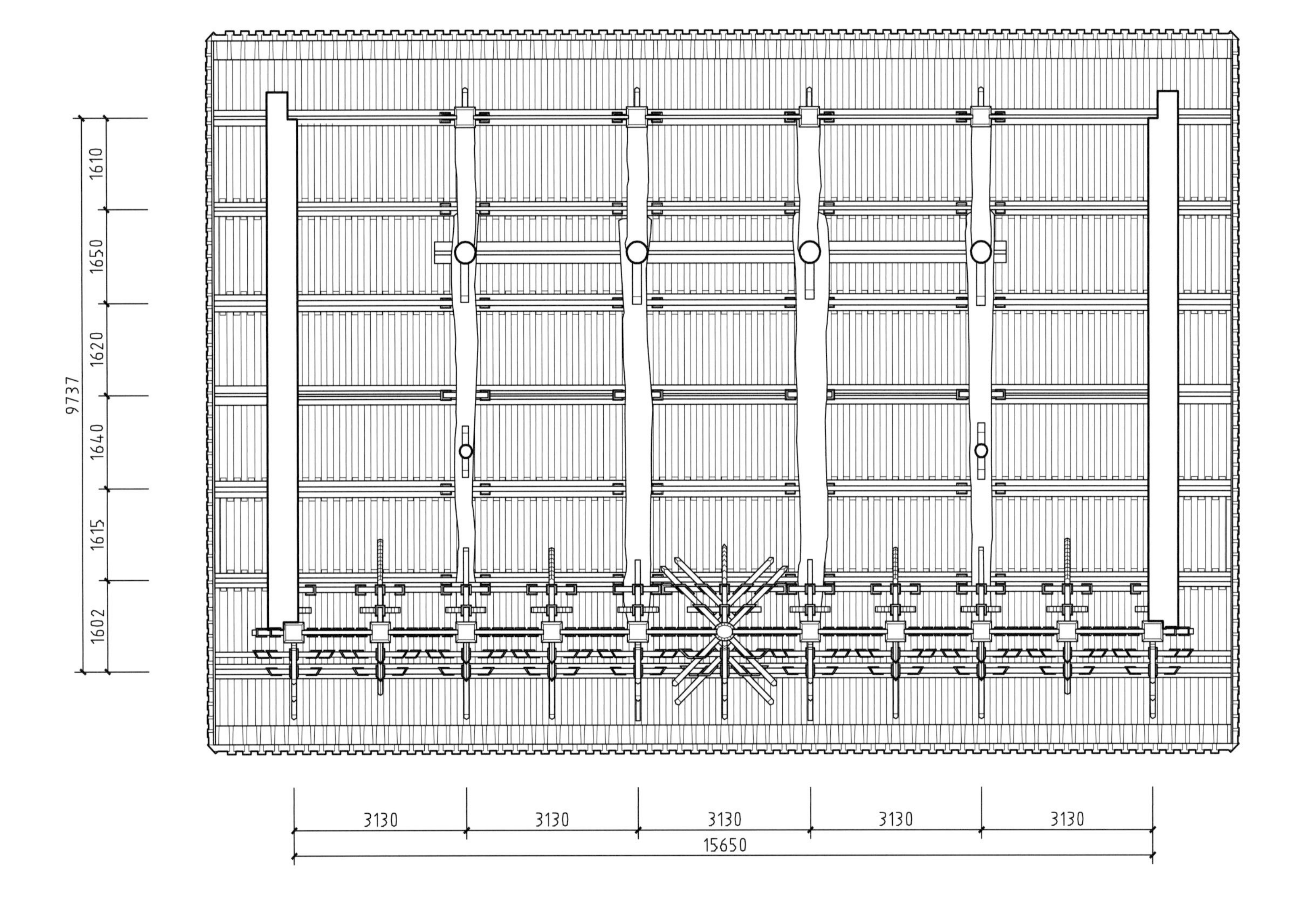

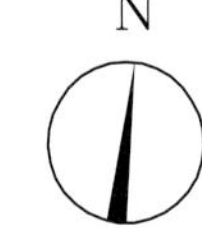

资圣寺后殿梁架仰视平面图
Plan of Zisheng Monastery's rear hall's framework as seen from below

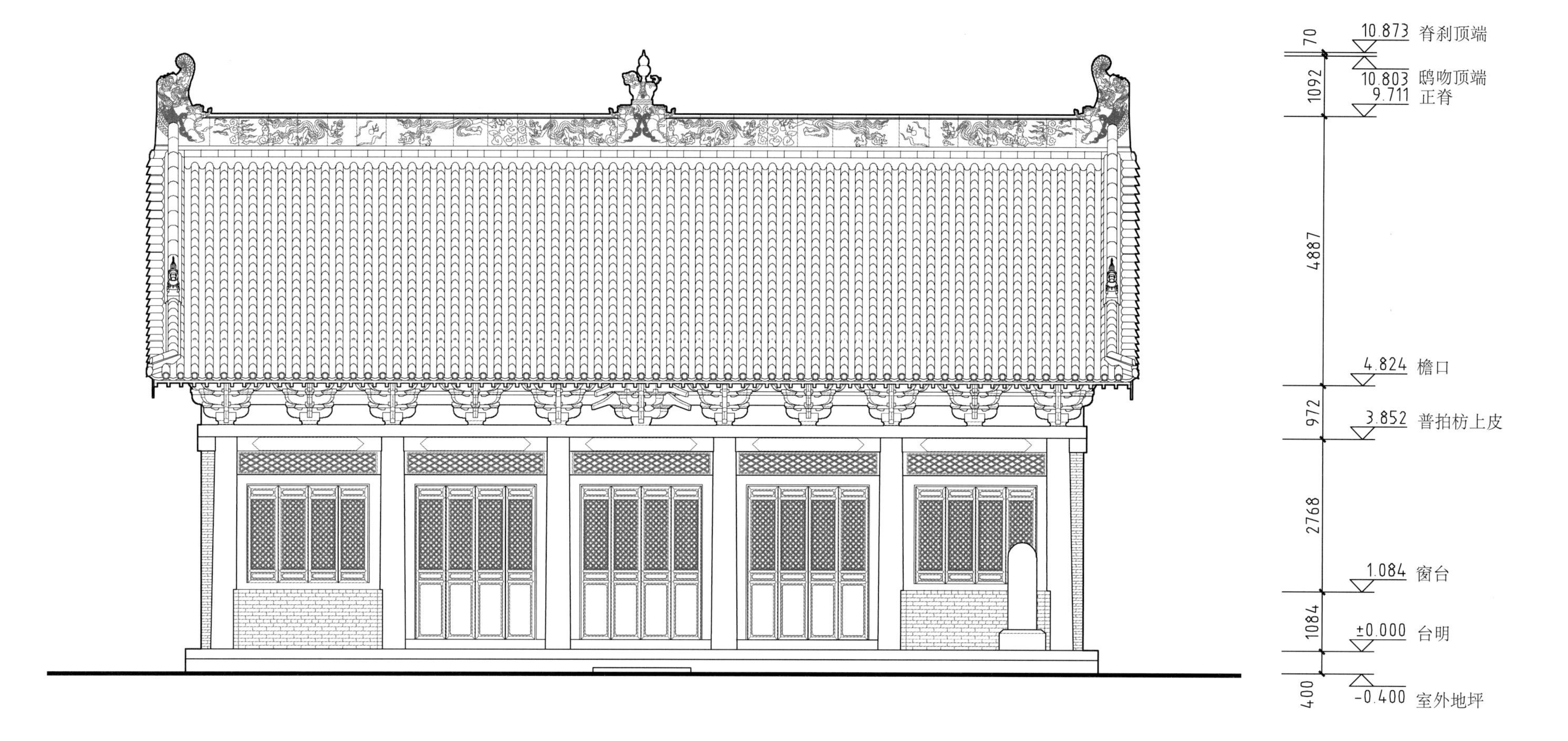

资圣寺后殿正立面图

Front elevation of Zisheng Monastery's rear hall

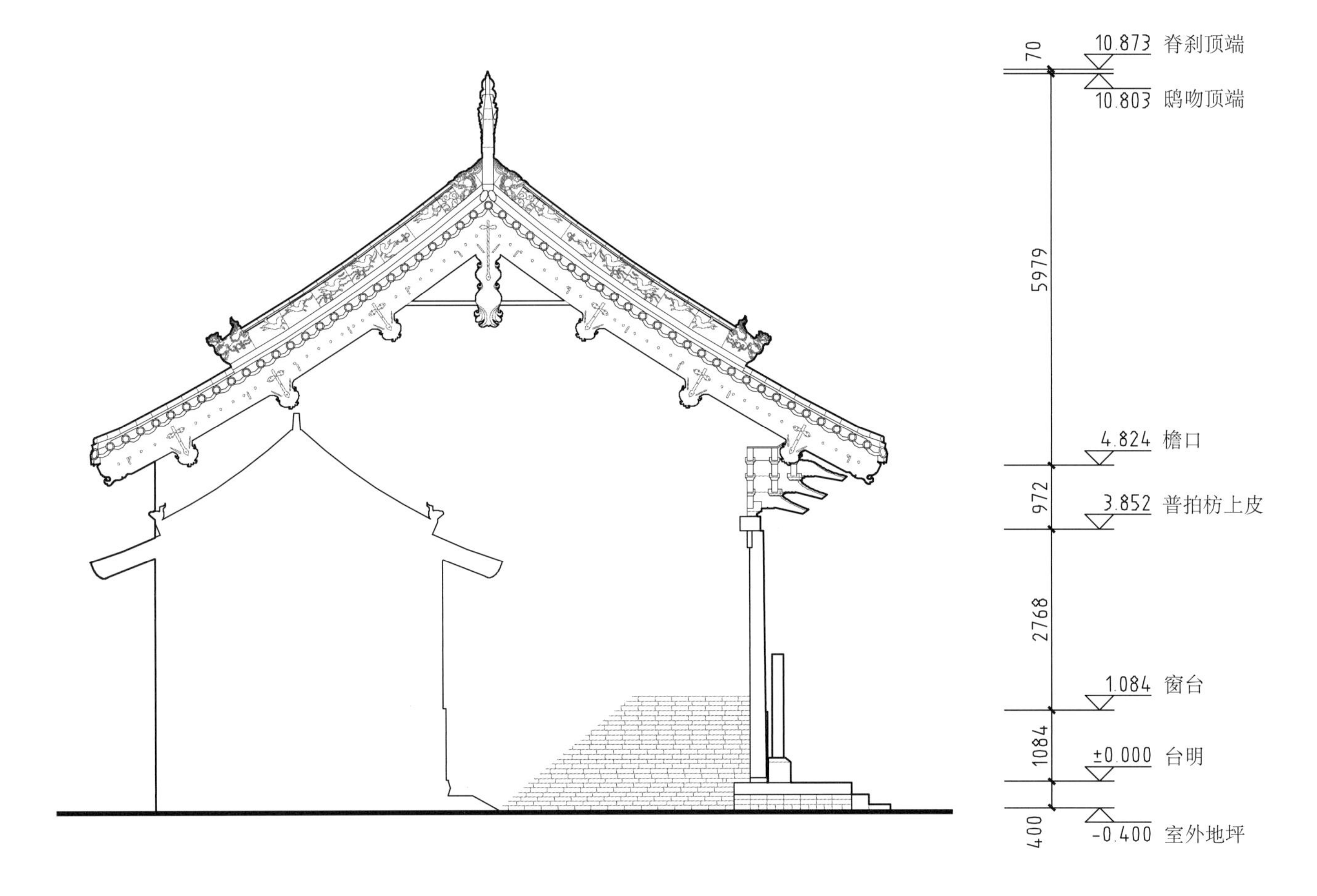

资圣寺后殿侧立面图

Side elevation of Zisheng Monastery's rear hall

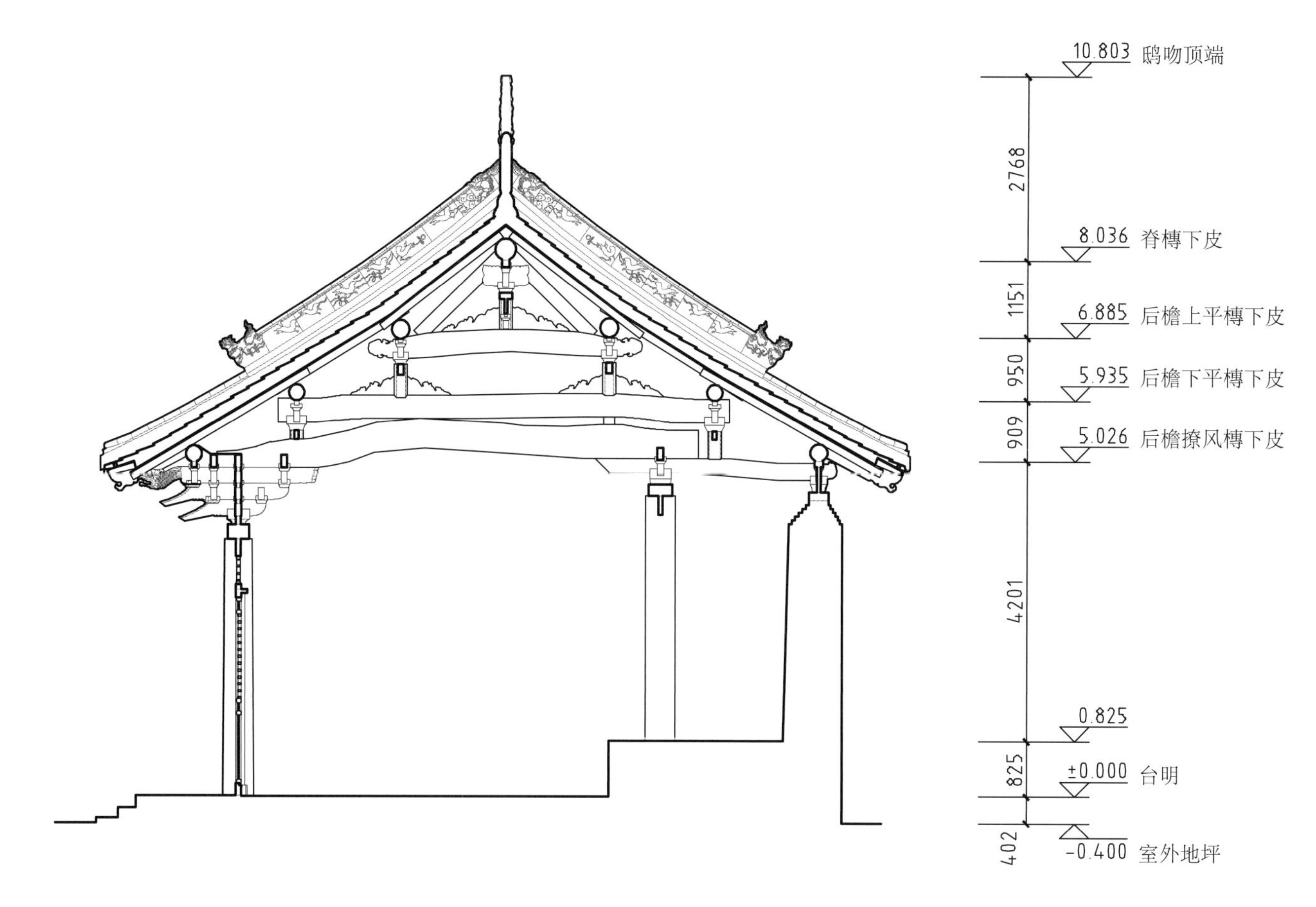

资圣寺后殿横剖面图
Cross-section of Zisheng Monastery's rear hall

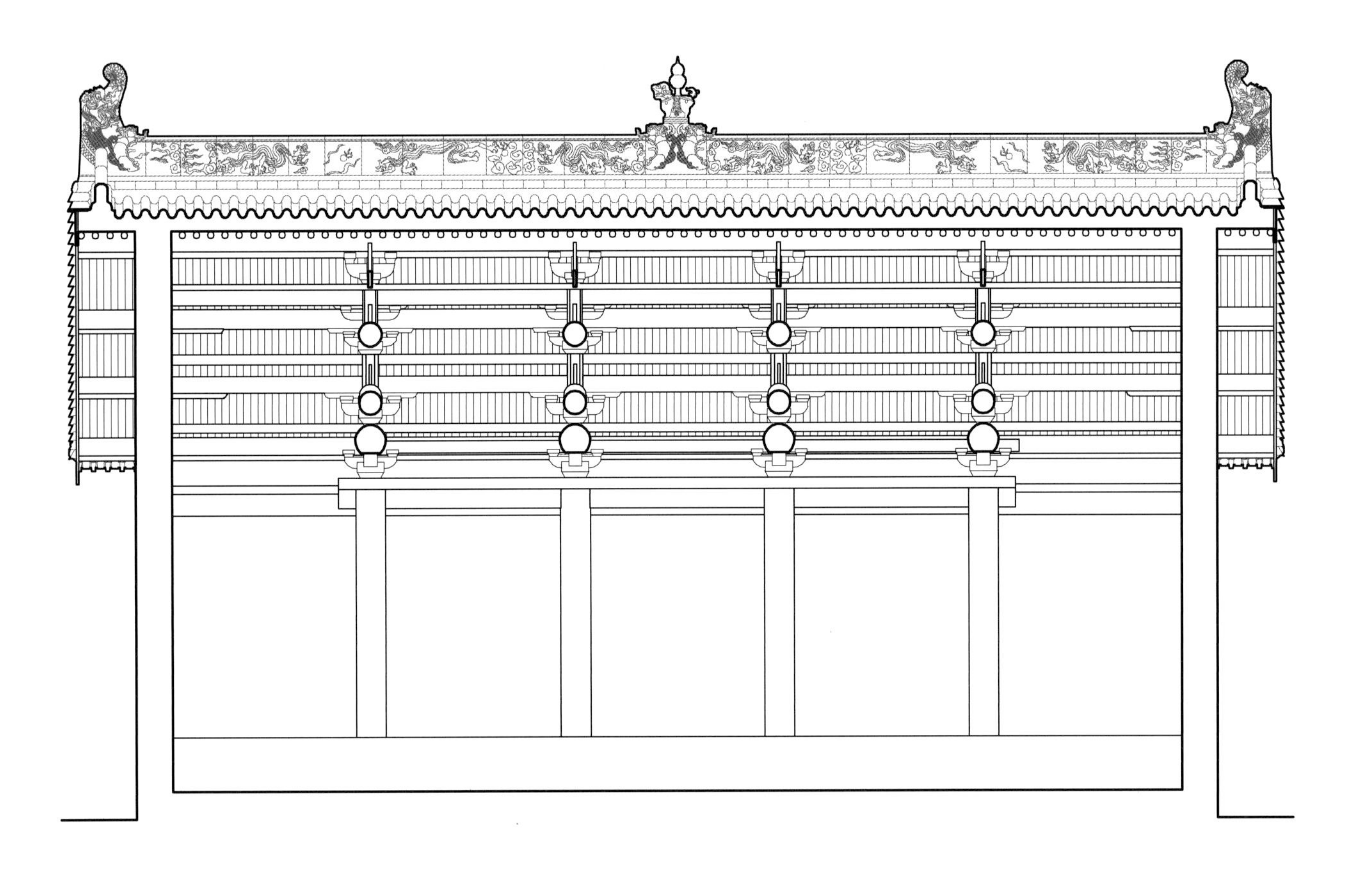

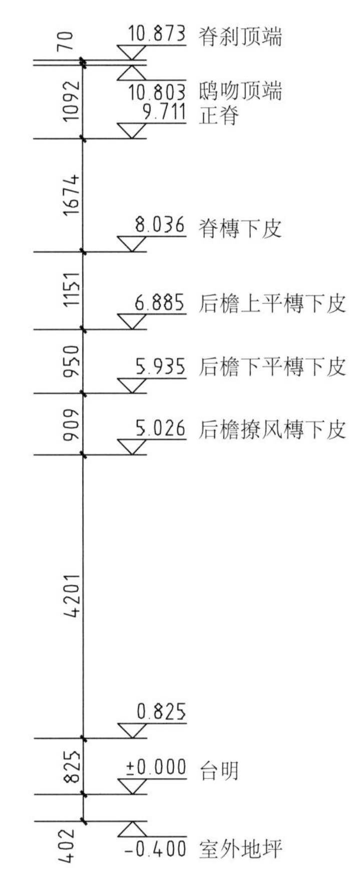

资圣寺后殿纵剖面图

Longitudinal section of Zisheng Monastery's rear hall

西李门二仙庙
Xilimen Erxian Temple

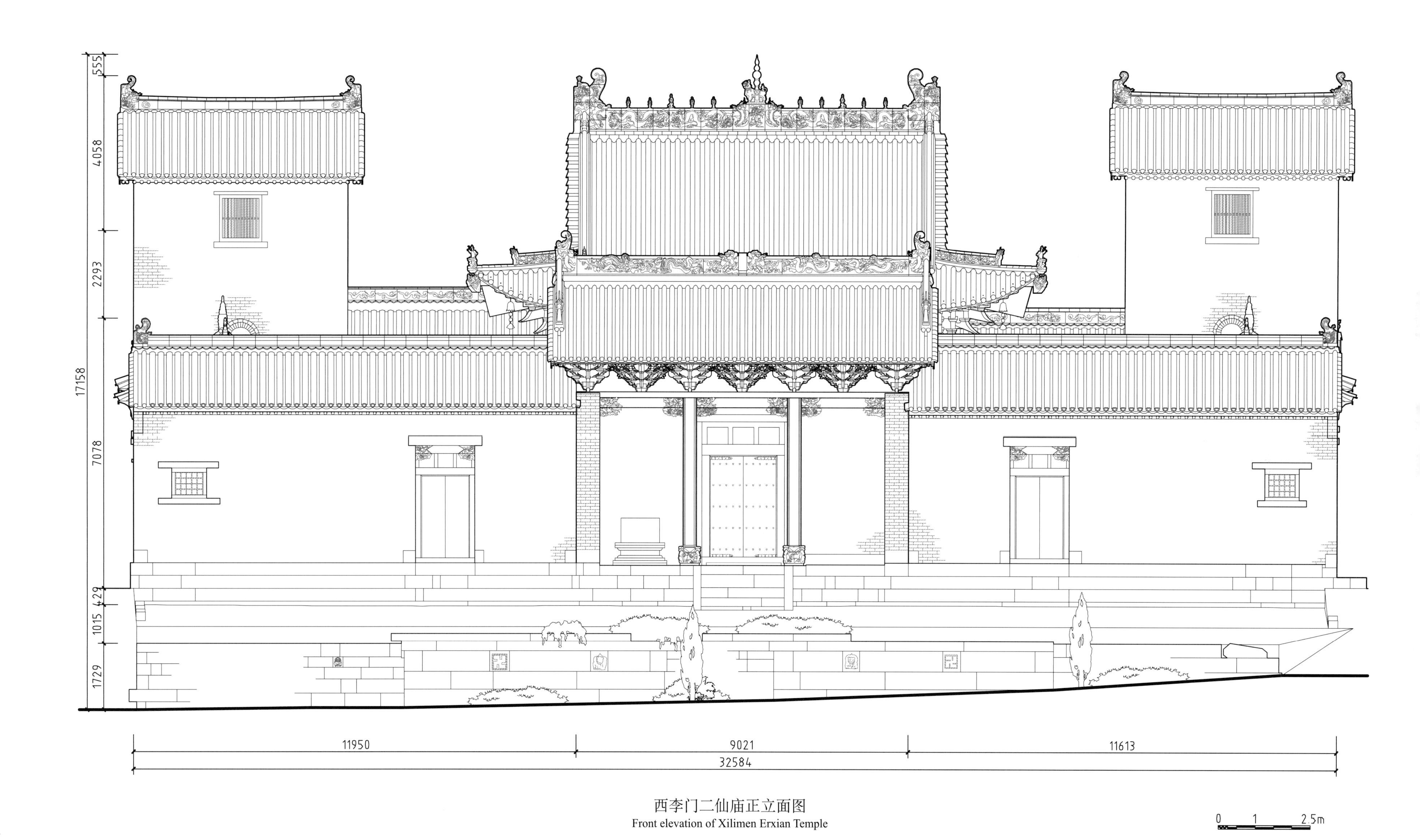

西李门二仙庙正立面图
Front elevation of Xilimen Erxian Temple

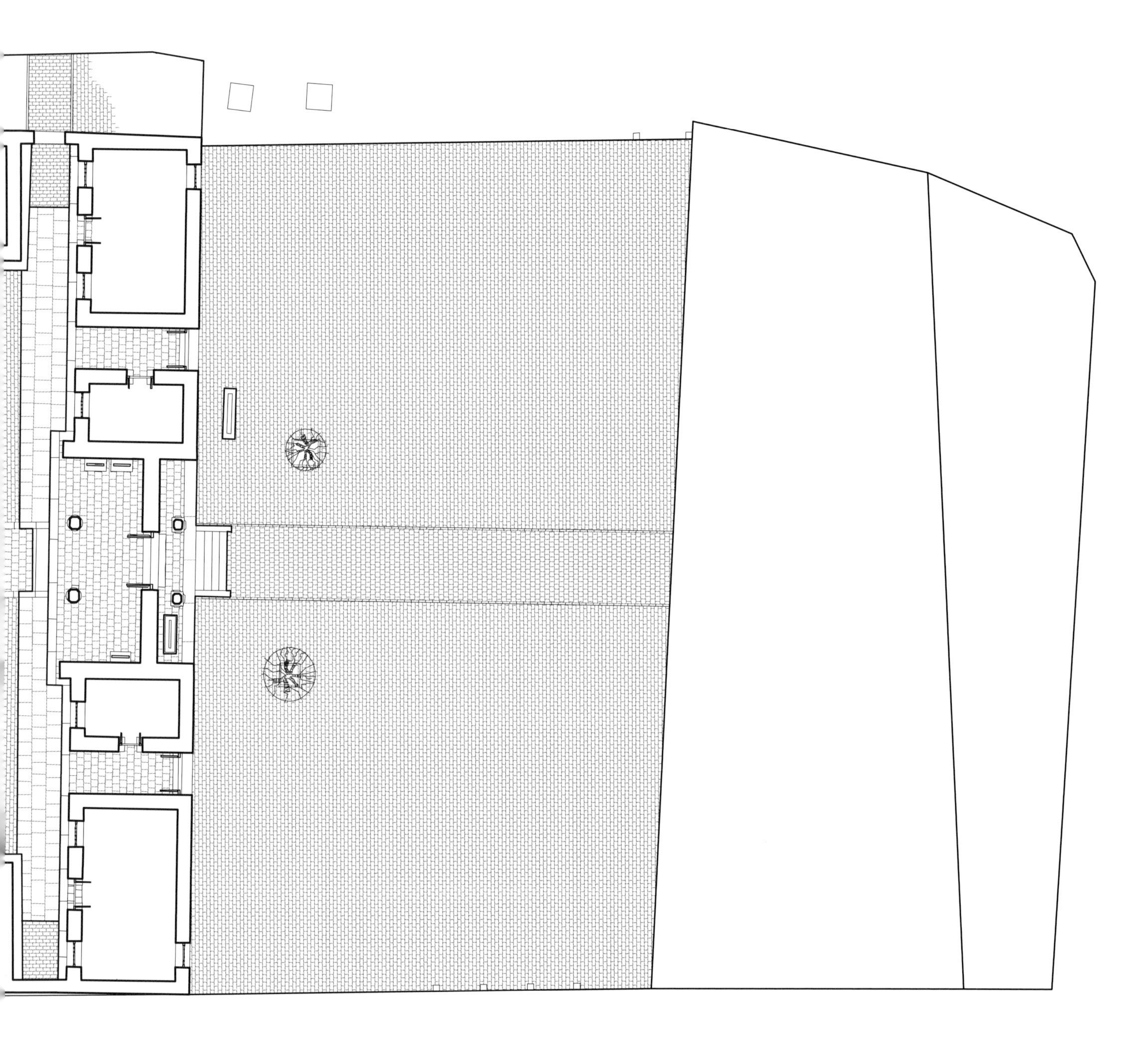

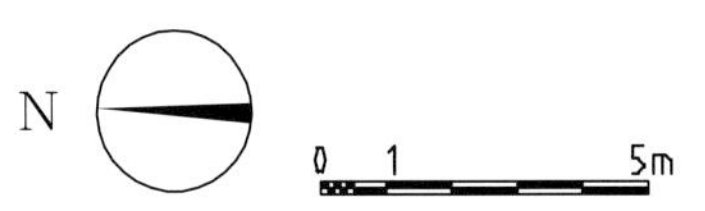
N
0 1 5m

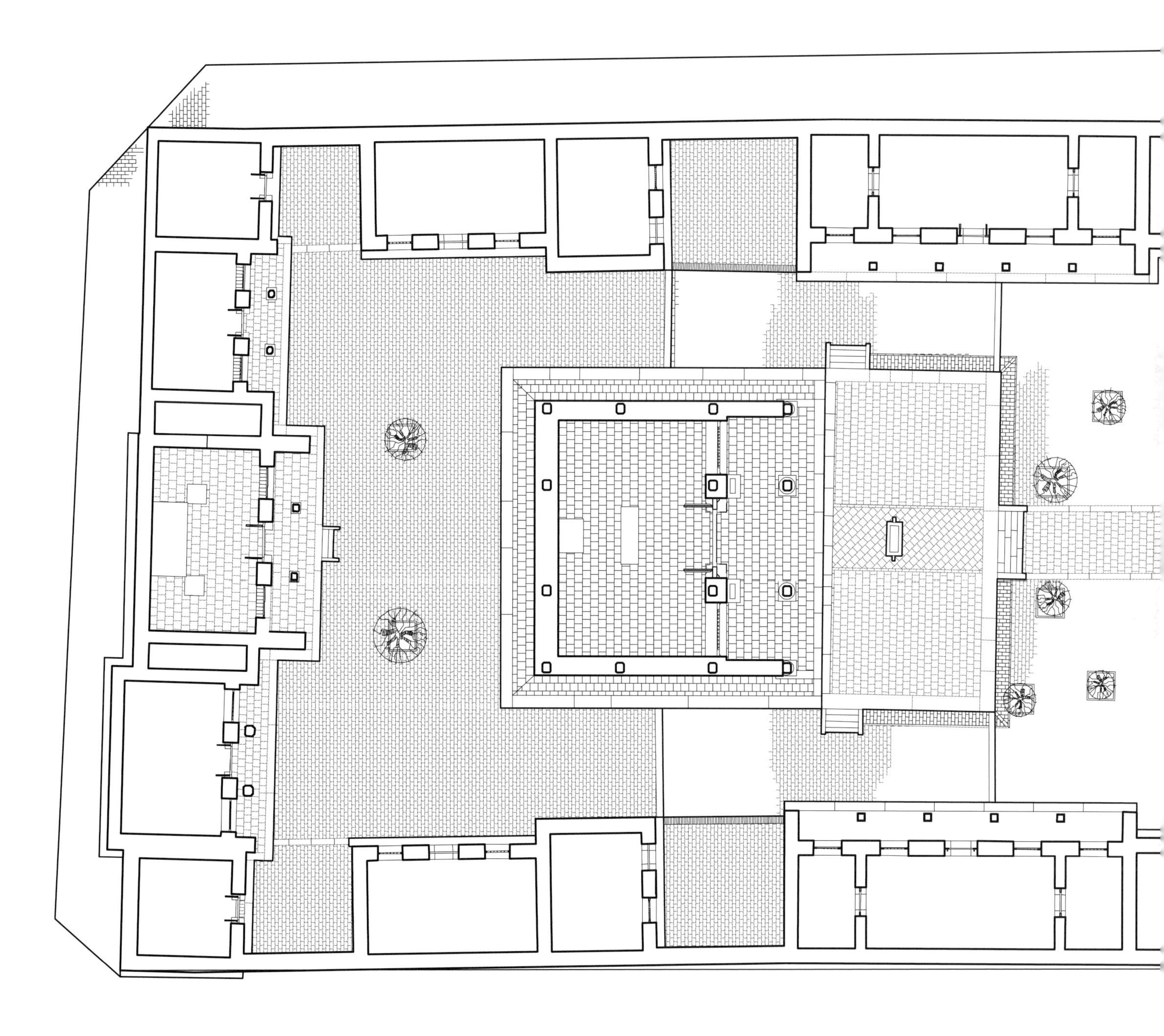

西李门二仙庙总平面图
Site plan of Xilimen Erxian Temple

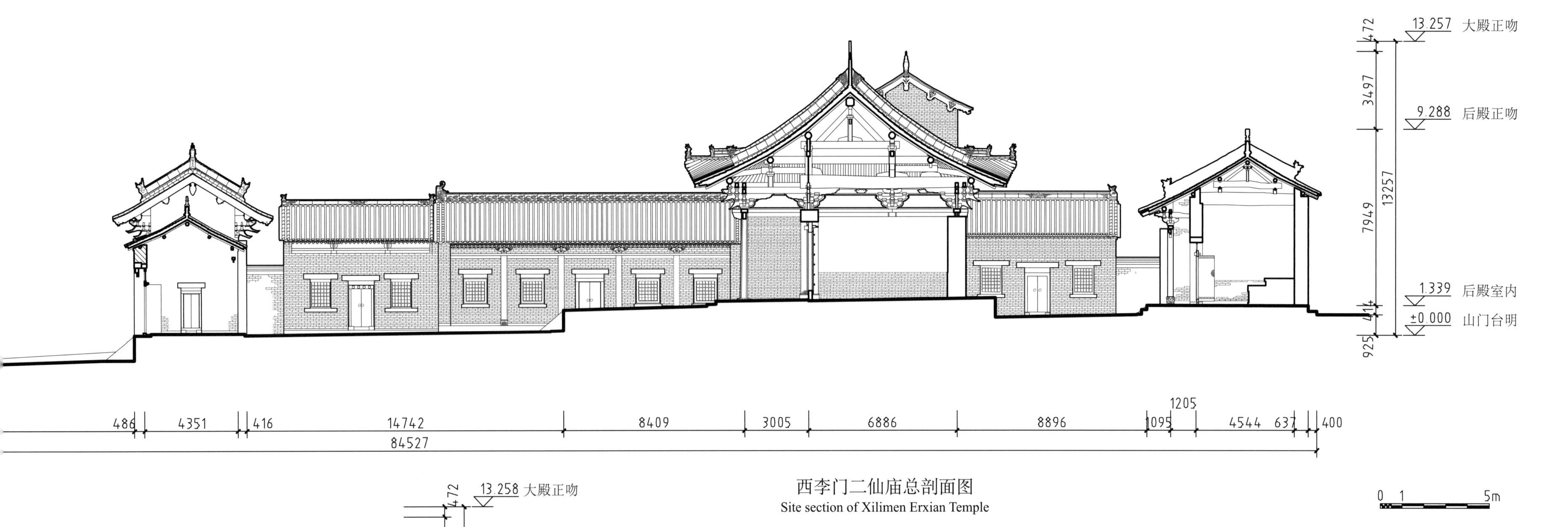

西李门二仙庙总剖面图
Site section of Xilimen Erxian Temple

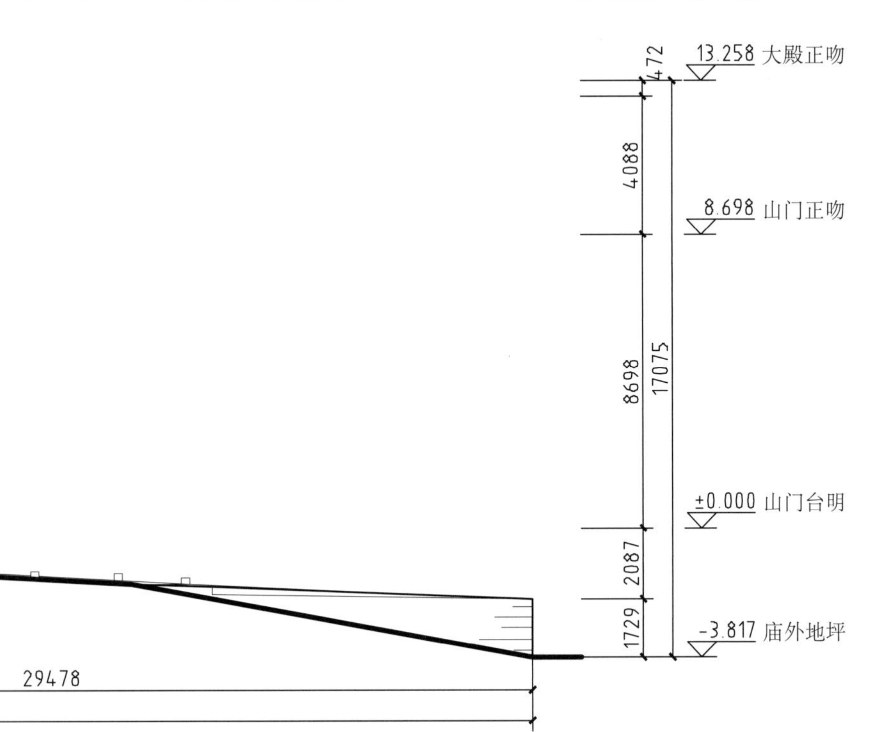

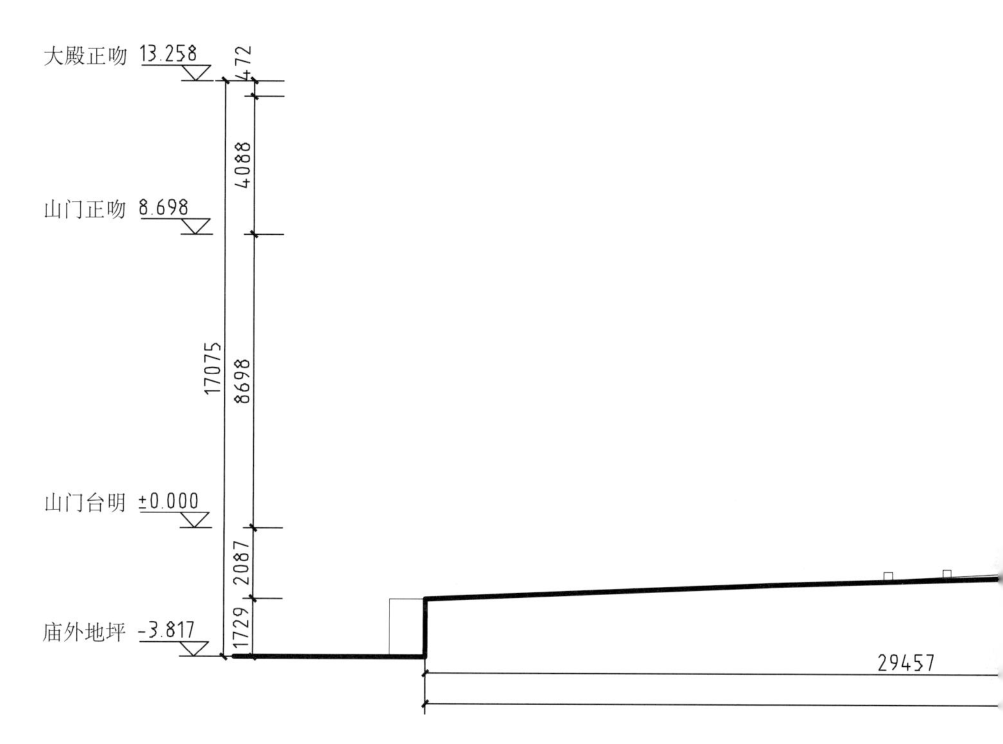

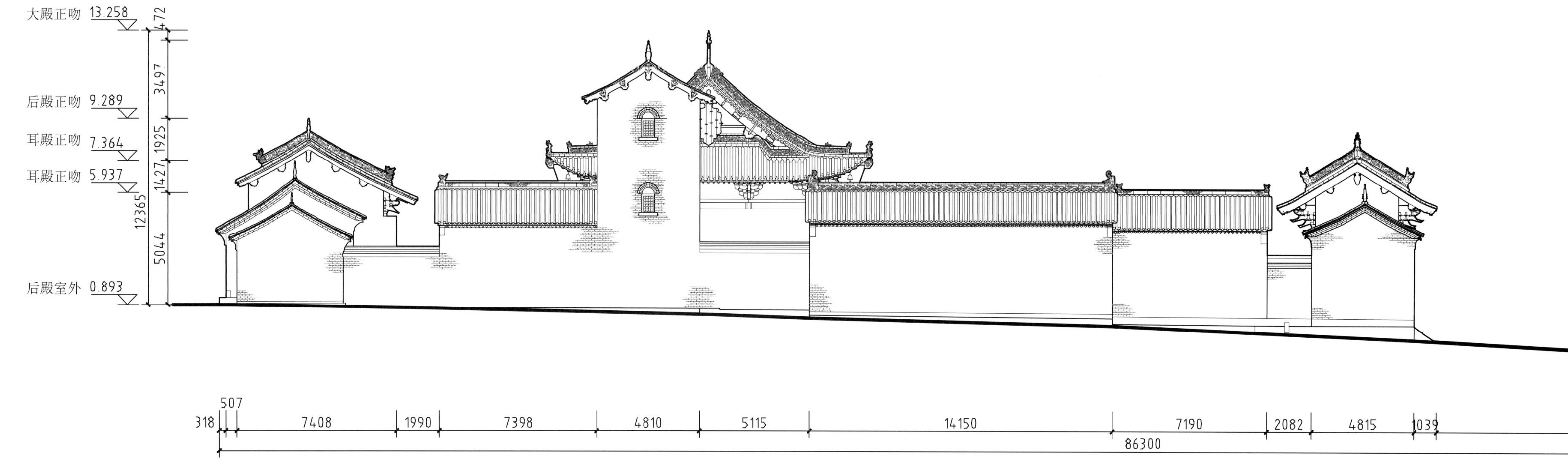

西李门二仙庙西立面图

West elevation of Xilimen Erxian Temple

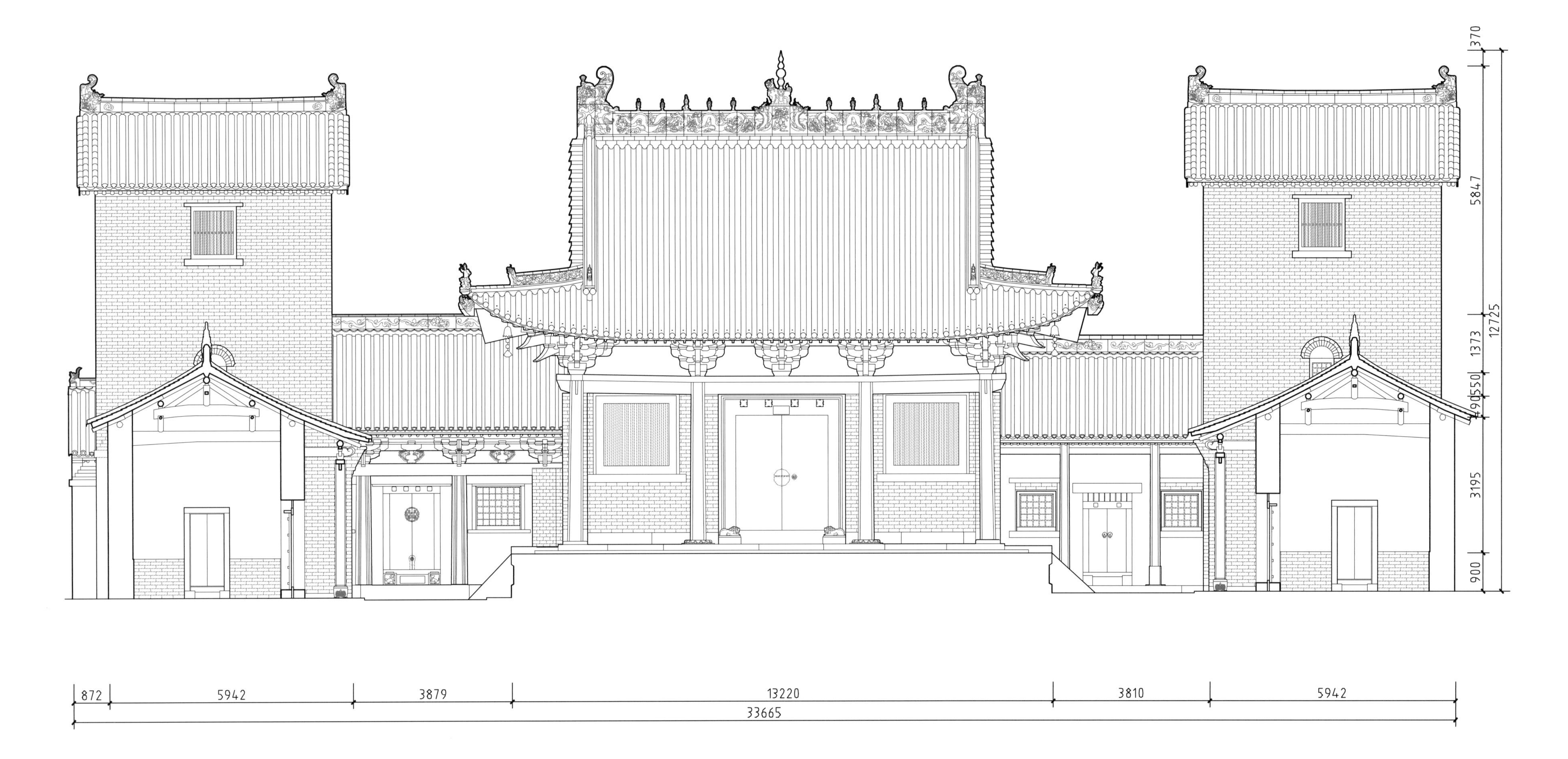

西李门二仙庙剖立面图
Sectional elevation of Xilimen Erxian Temple

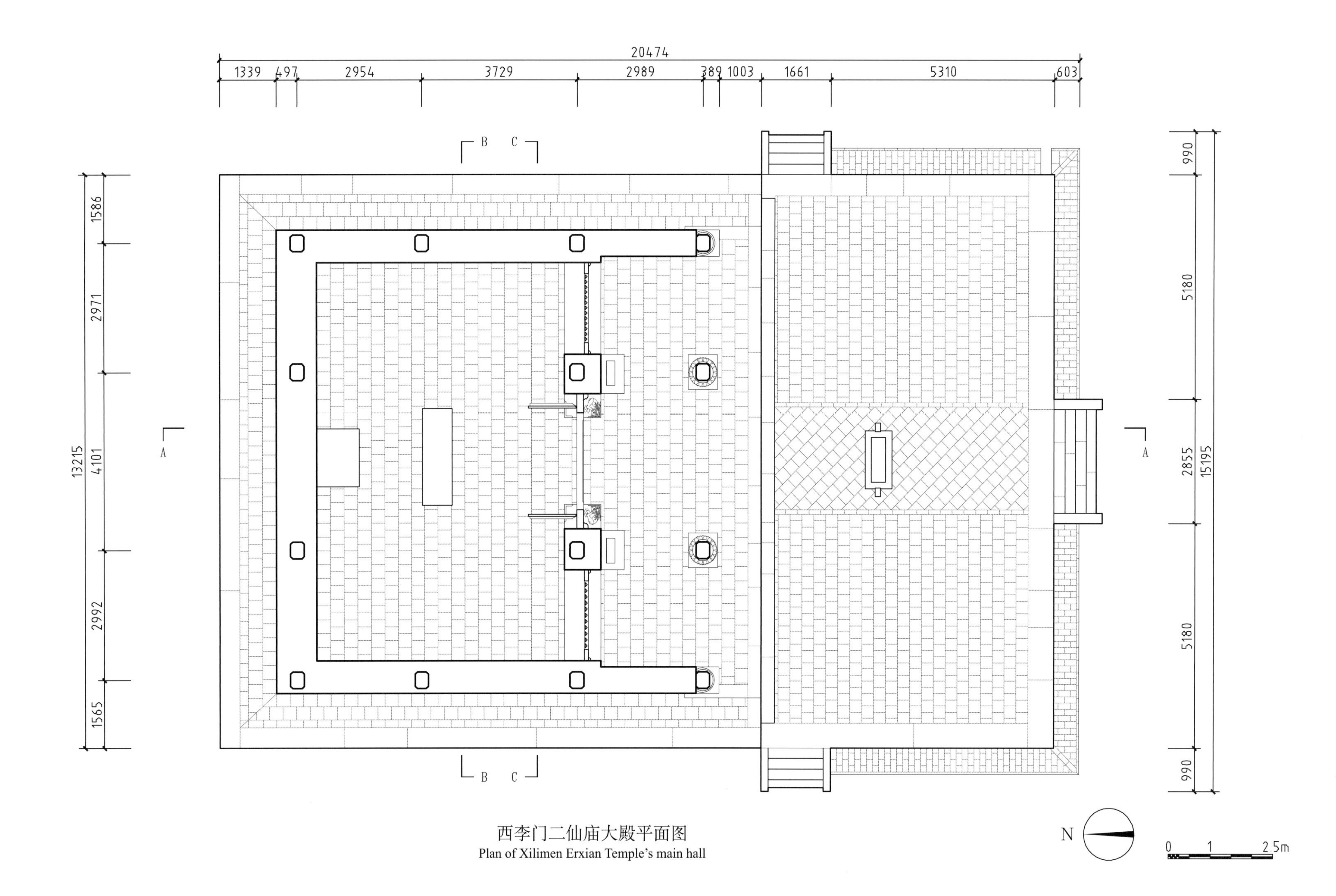

西李门二仙庙大殿平面图

Plan of Xilimen Erxian Temple's main hall

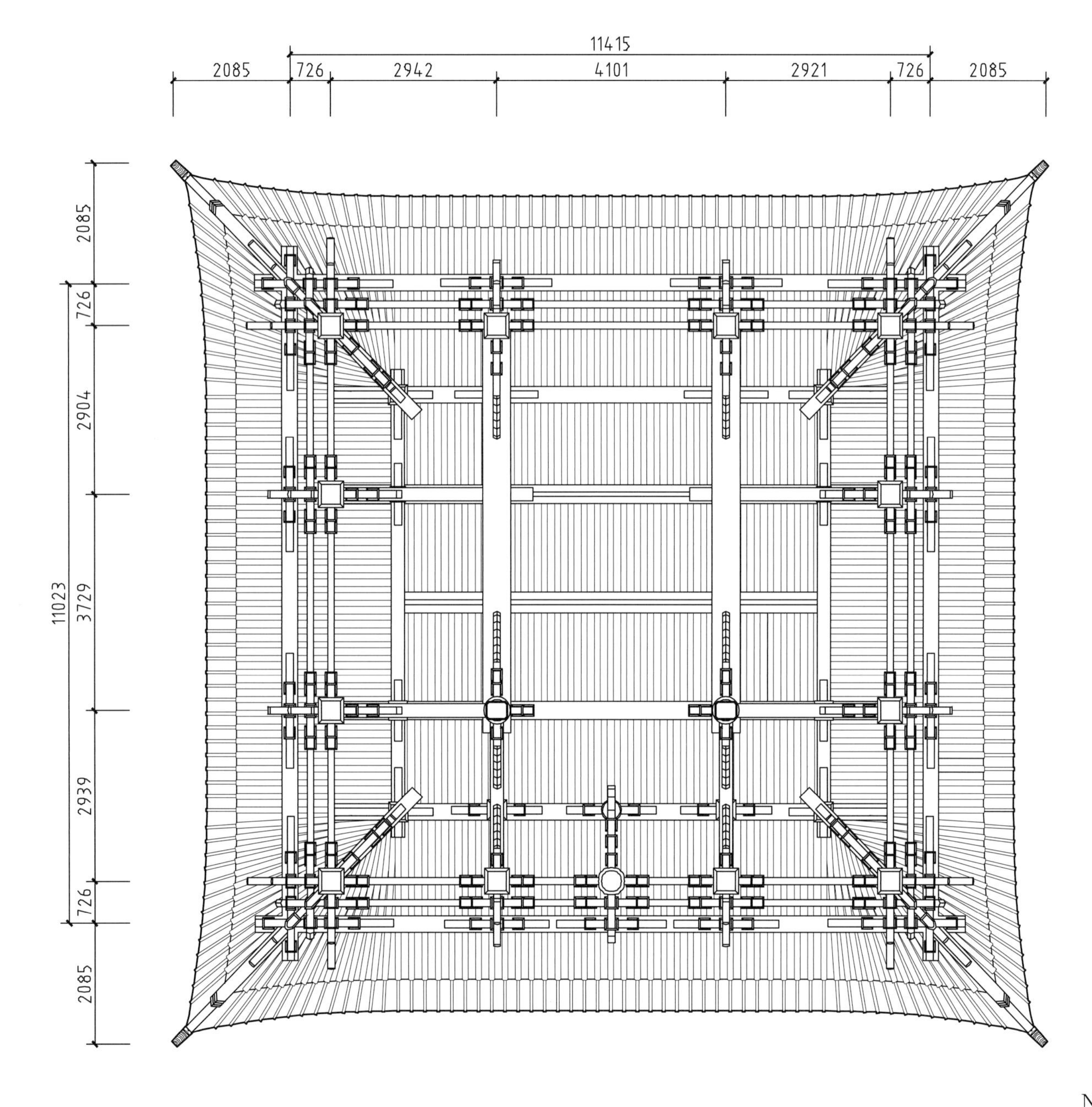

N

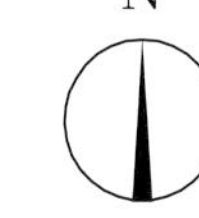

0 1 2.5m

西李门二仙庙大殿梁架仰视图

Plan of Xilimen Erxian Temple's main hall's framework as seen from below

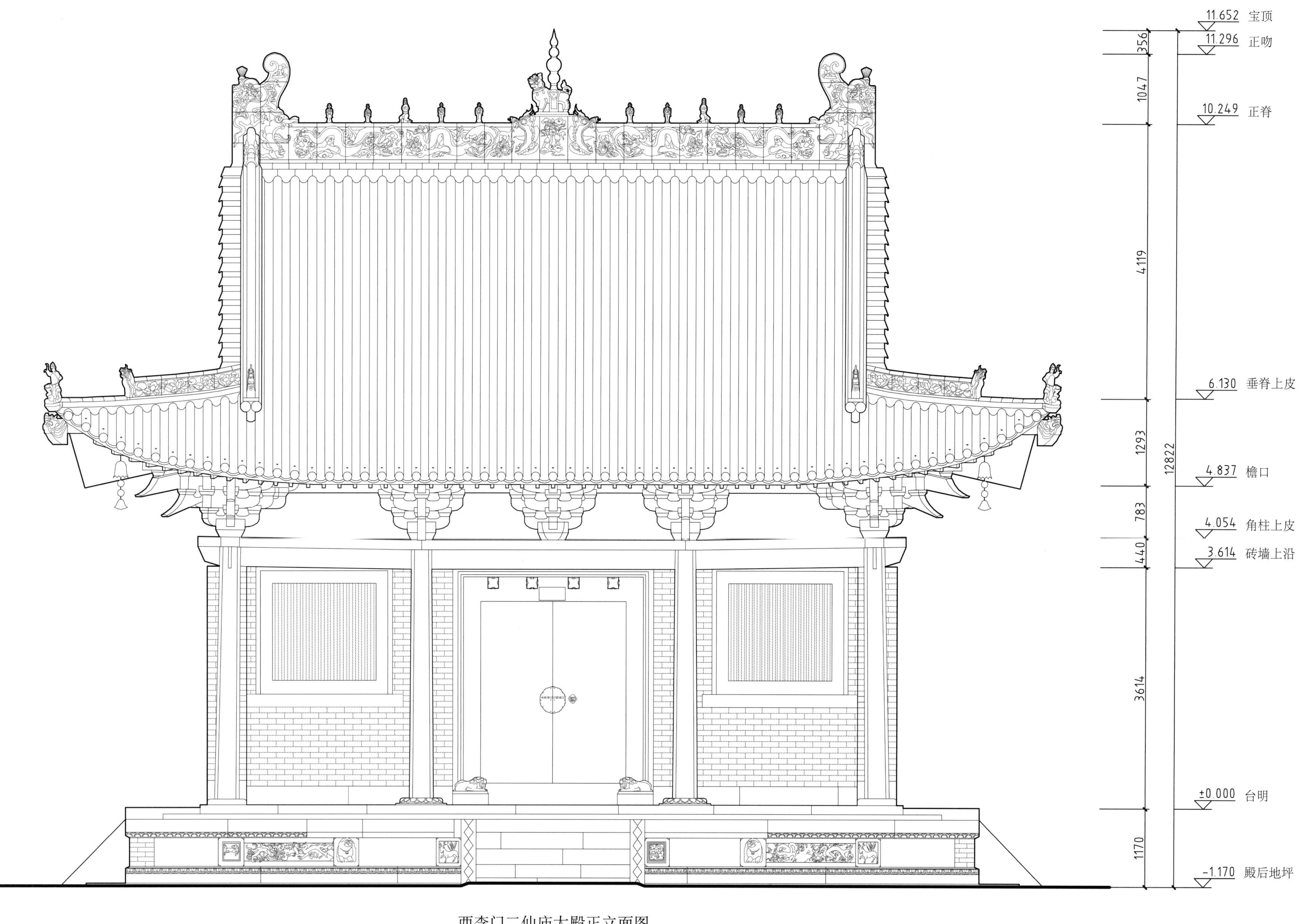

西李门二仙庙大殿正立面图

Front elevation of Xilimen Erxian Temple's main hall

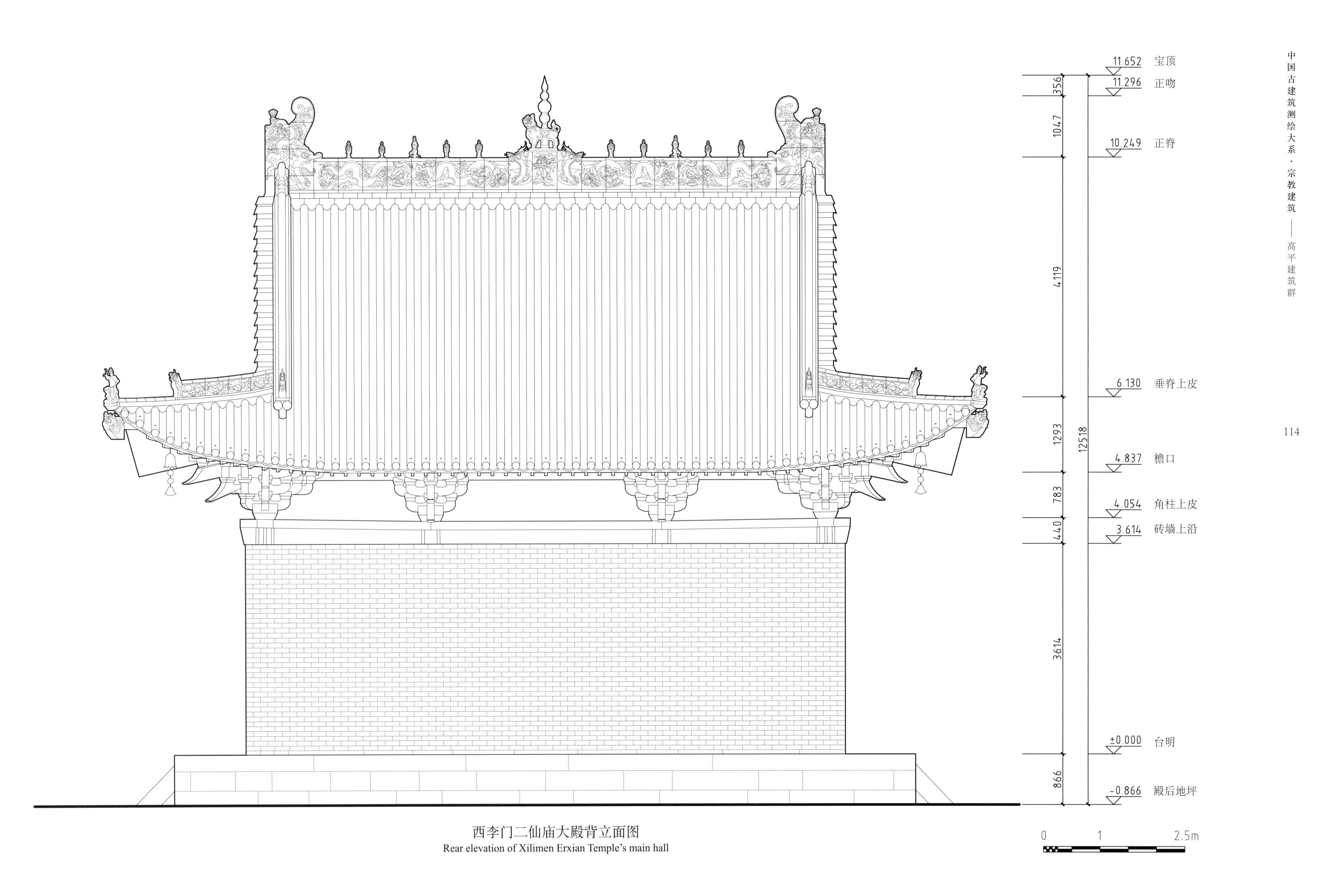

西李门二仙庙大殿背立面图

Rear elevation of Xilimen Erxian Temple's main hall

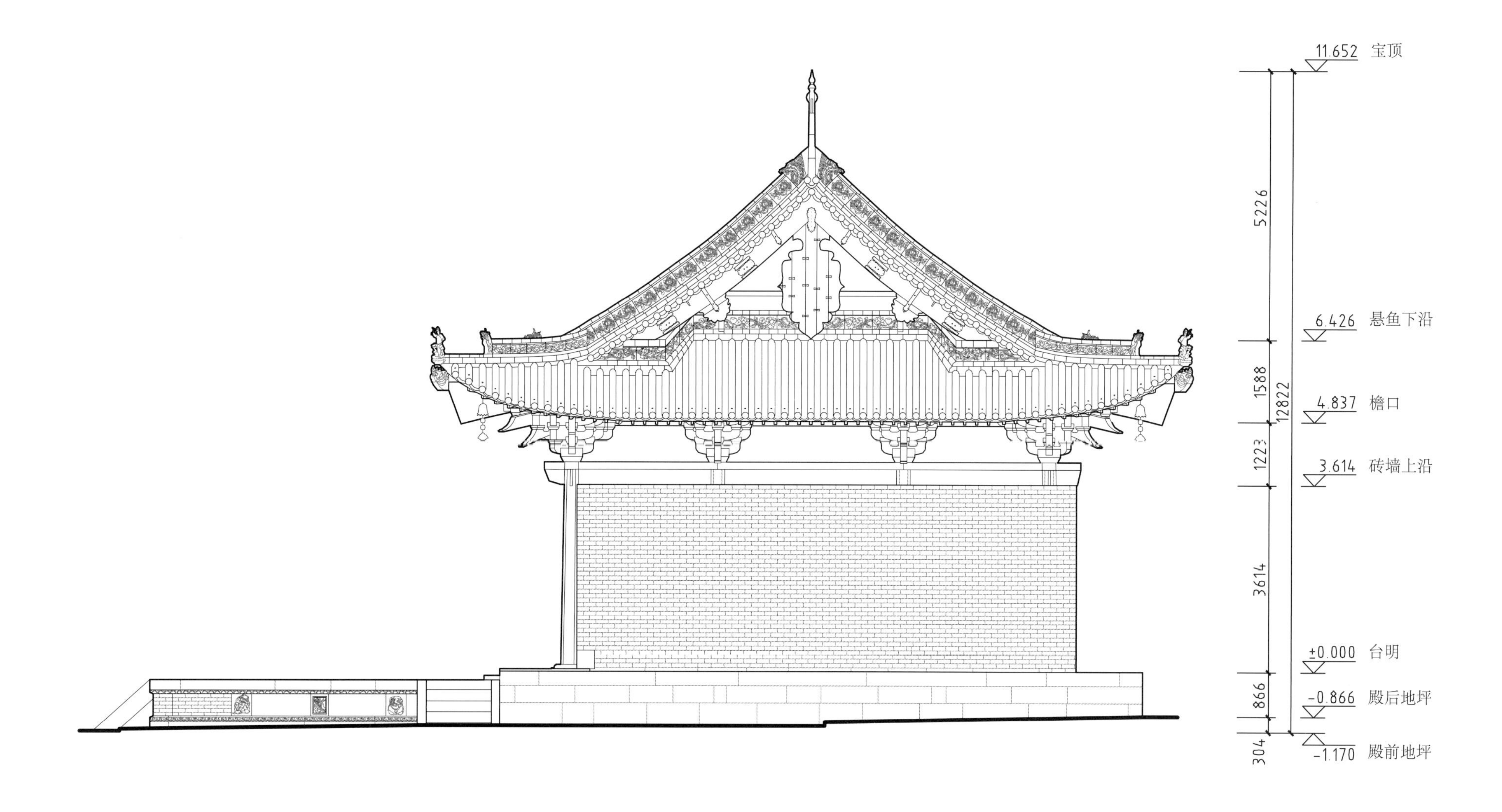

西李门二仙庙大殿侧立面图

Side elevation of Xilimen Erxian Temple's main hall

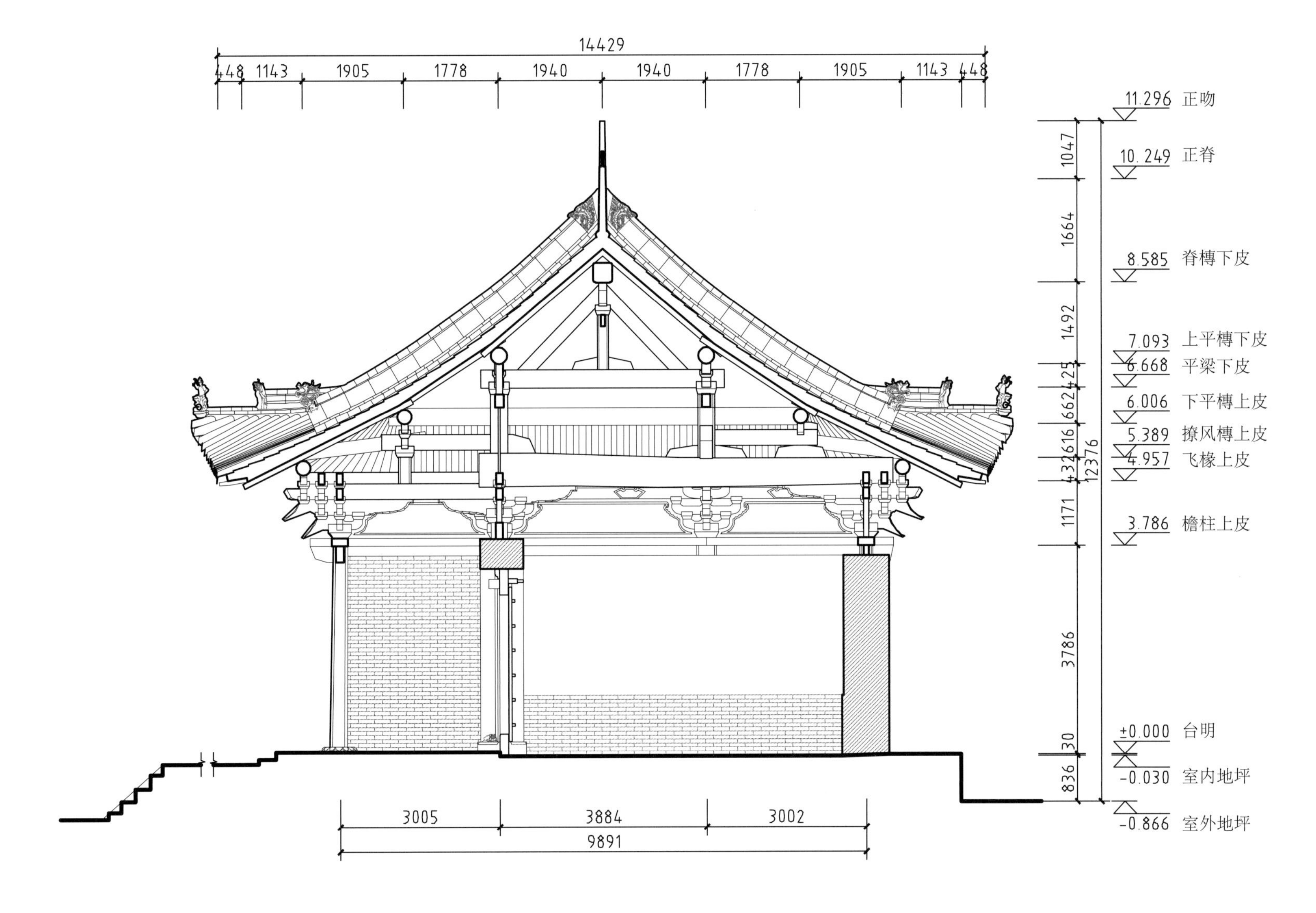

西李门二仙庙大殿 A-A 剖面图
Section A-A of Xilimen Erxian Temple's main hall

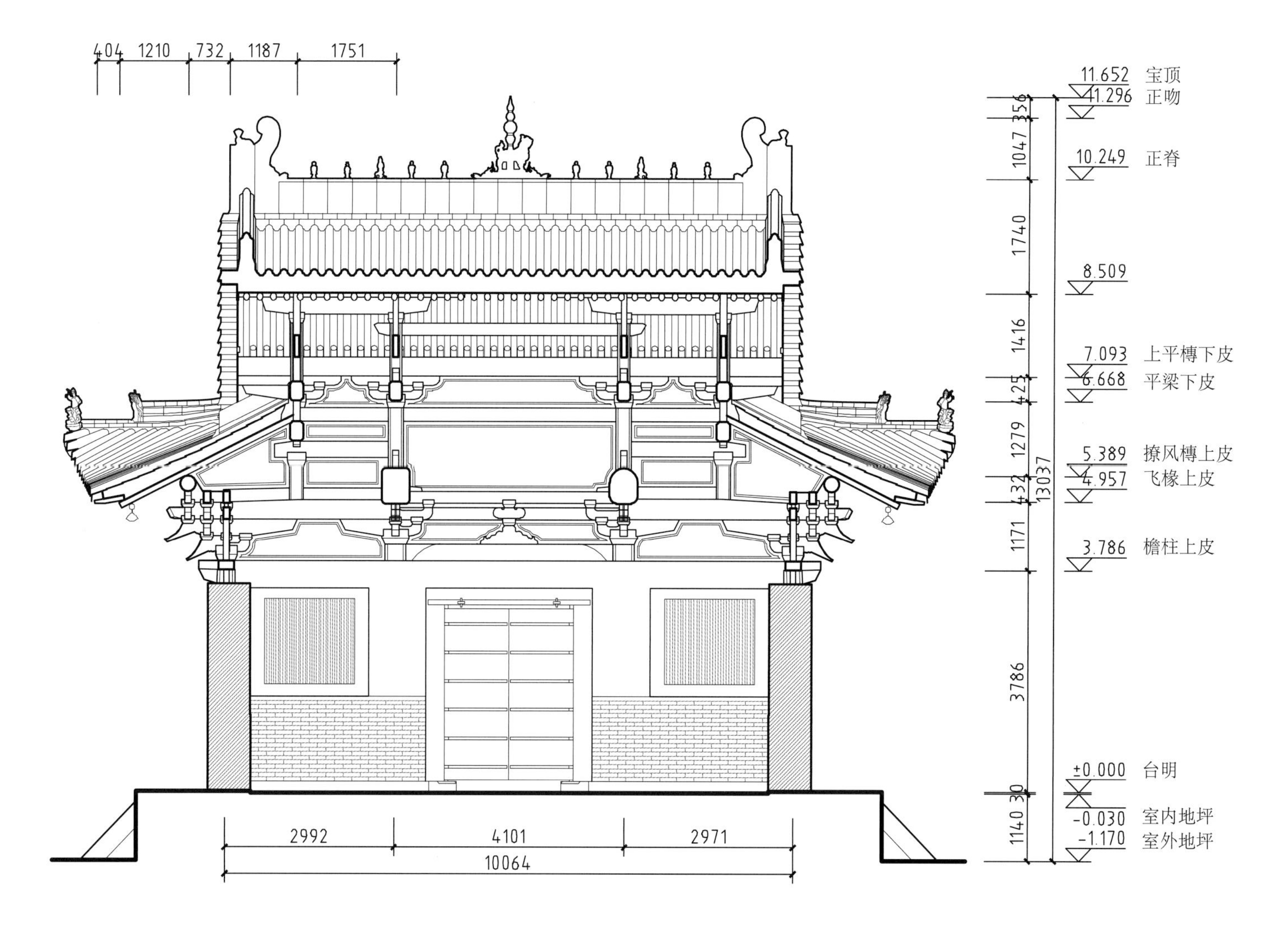

西李门二仙庙大殿 B—B 剖面图
Section B-B of Xilimen Erxian Temple's main hall

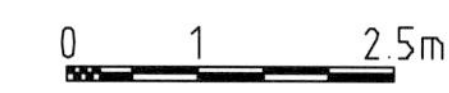

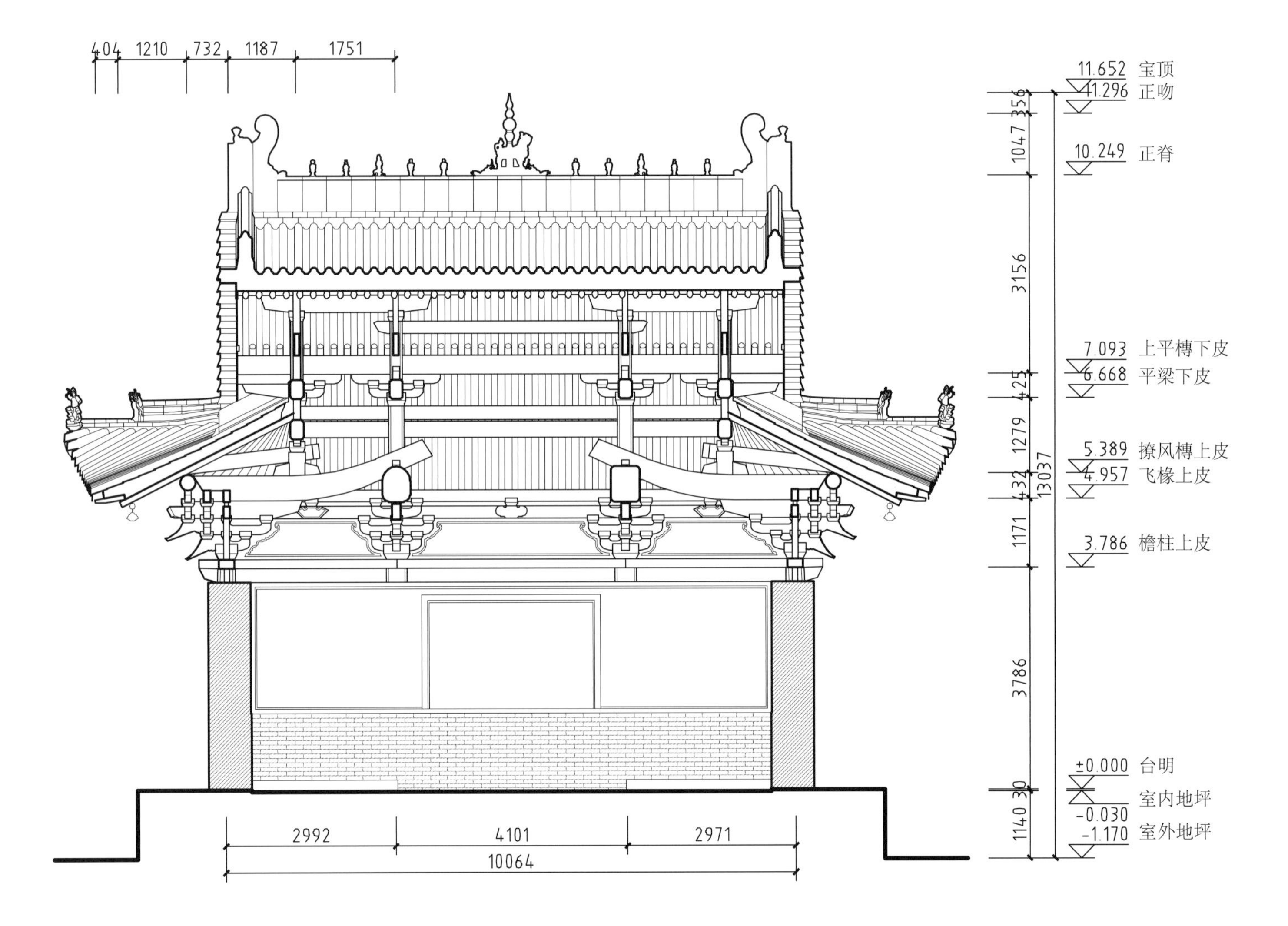

西李门二仙庙大殿 C-C 剖面图
Section C-C of Xilimen Erxian Temple's main hall

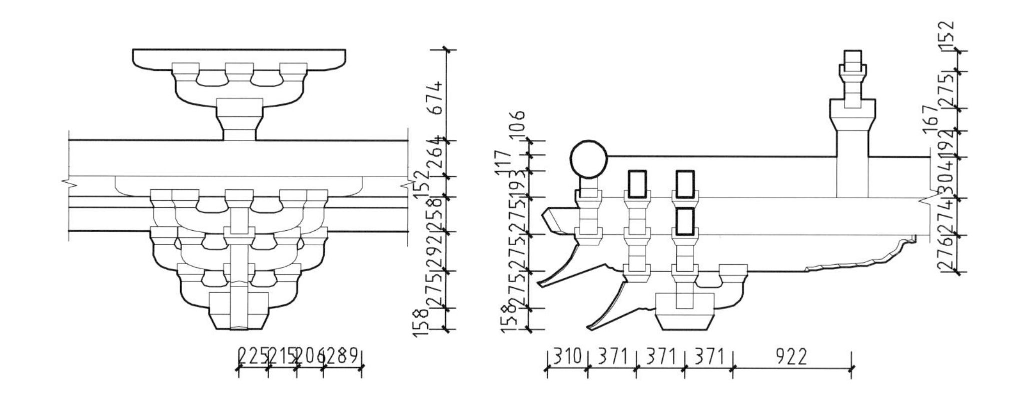

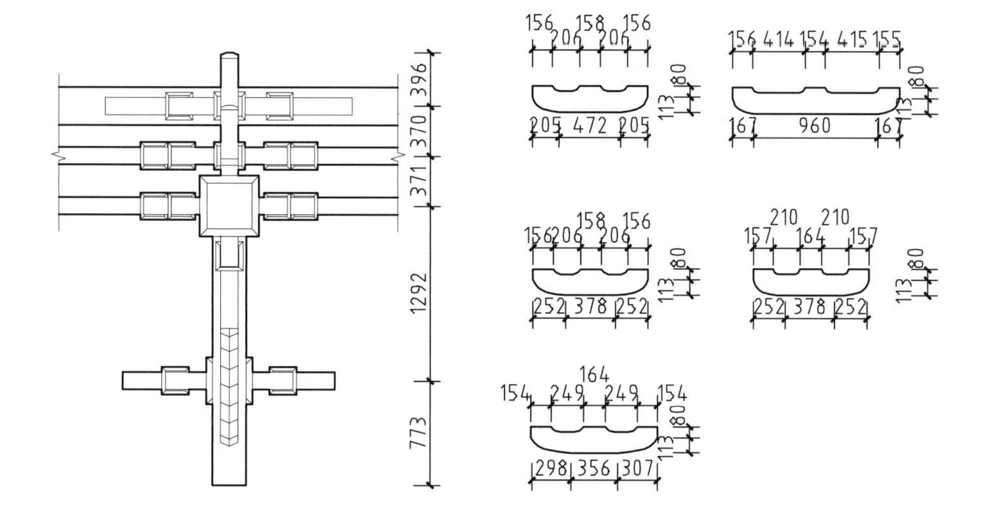

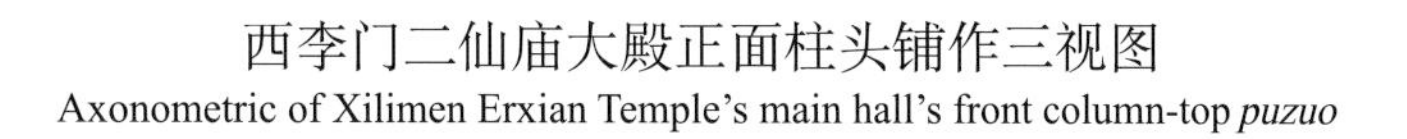

西李门二仙庙大殿正面柱头铺作三视图

Axonometric of Xilimen Erxian Temple's main hall's front column-top *puzuo*

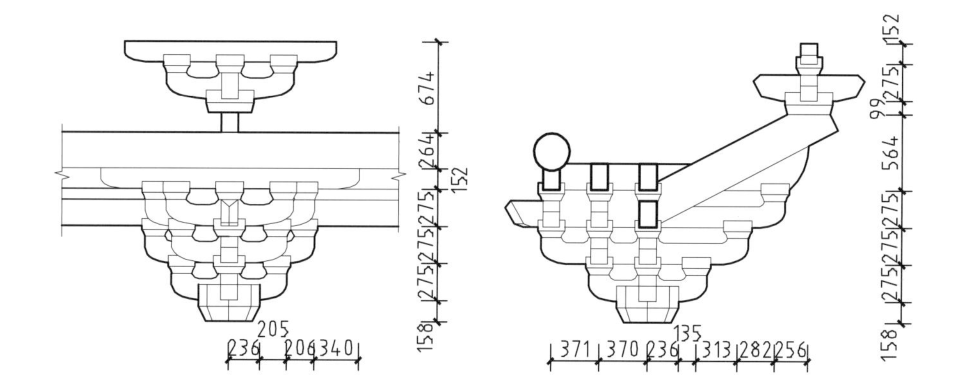

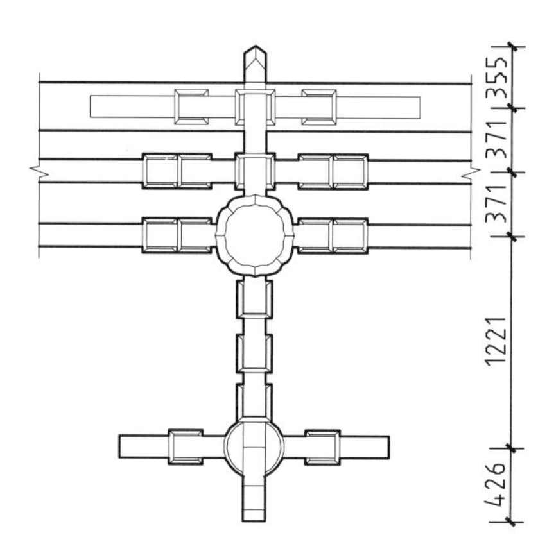

西李门二仙庙大殿正面补间铺作三视图

Axonometric of Xilimen Erxian Temple's main hall's intercolumnar bracket set

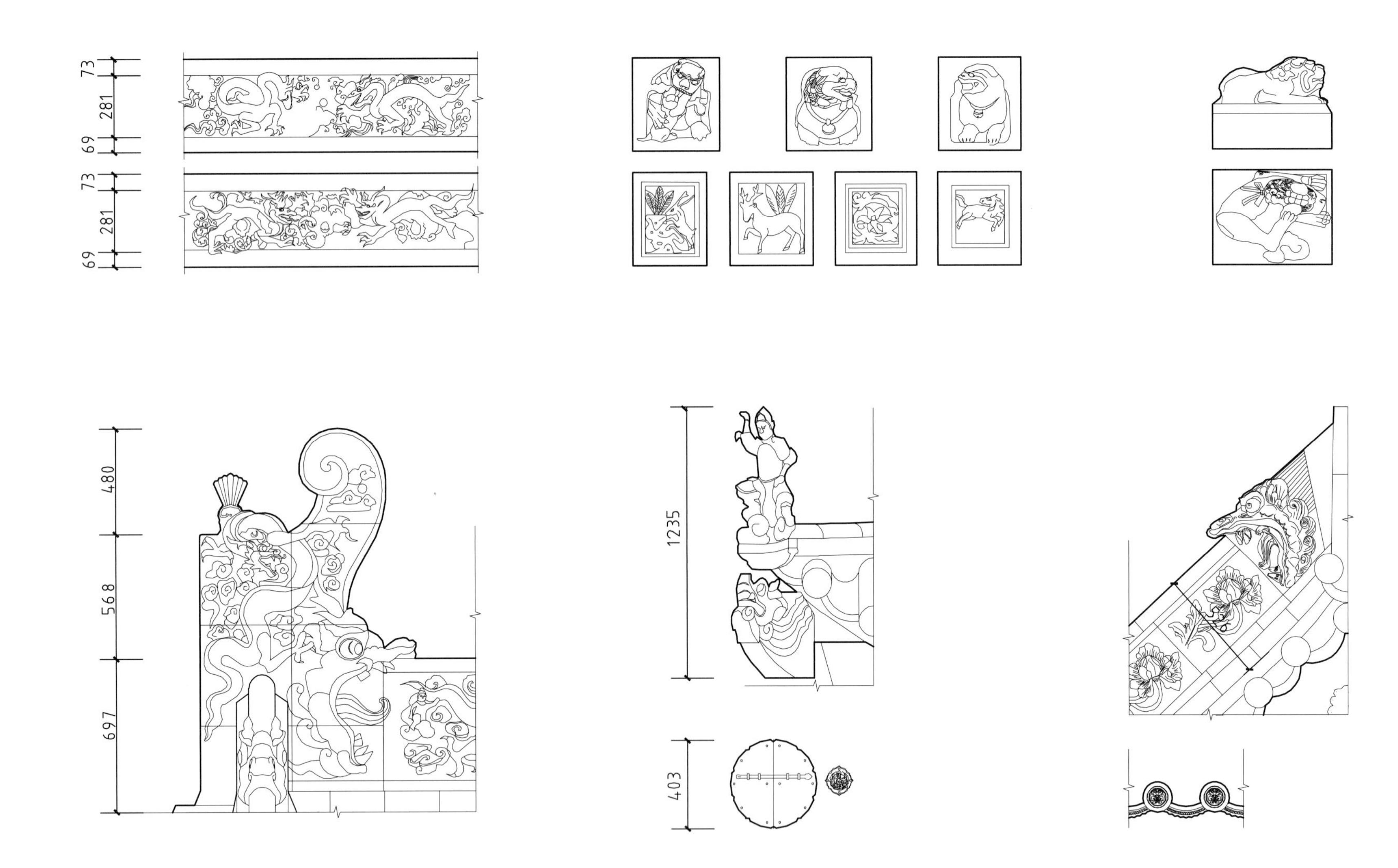

西李门二仙庙大殿立面细部大样图

Elevation detail of Xilimen Erxian Temple's main hall

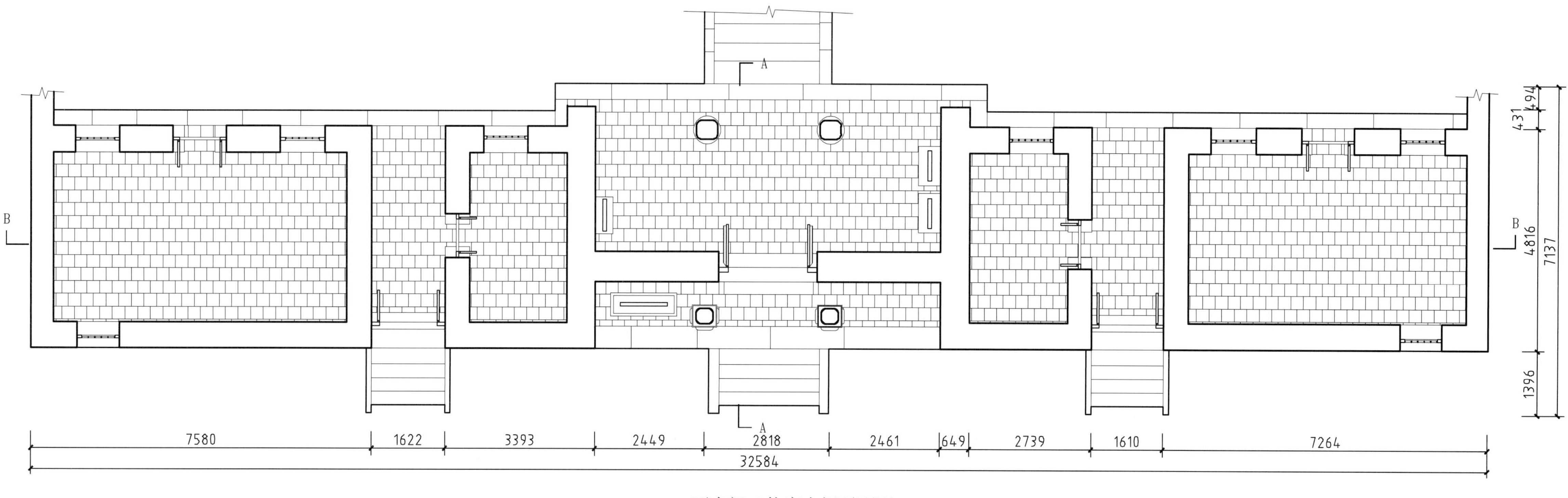

西李门二仙庙山门平面图
Plan of Xilimen Erxian Temple's *shanmen*

0 1 2.5m

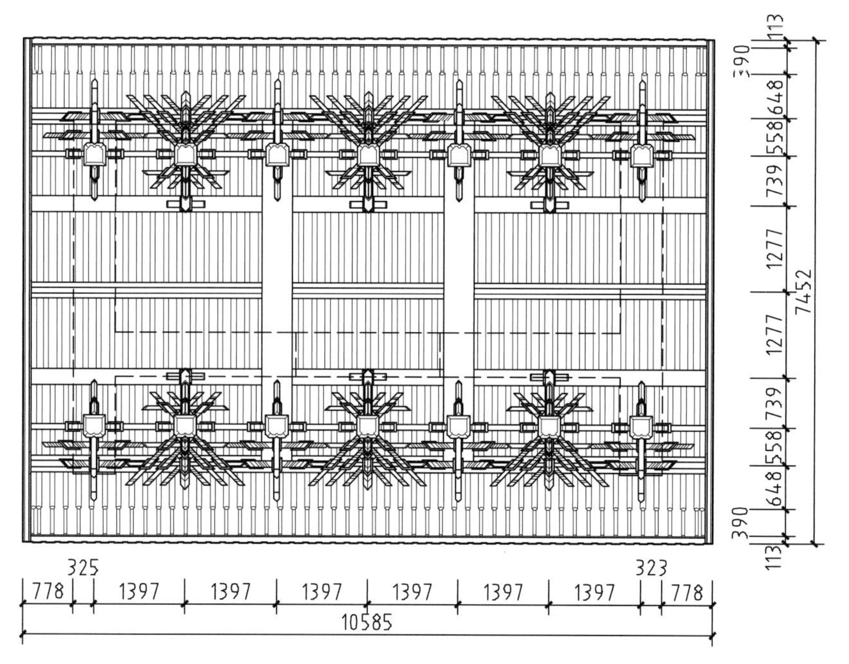

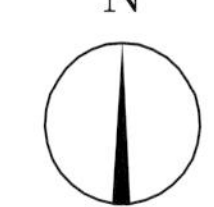

西李门二仙庙山门梁架仰视平面图
Plan of Xilimen Erxian Temple's *shanmen*'s framework as seen from below

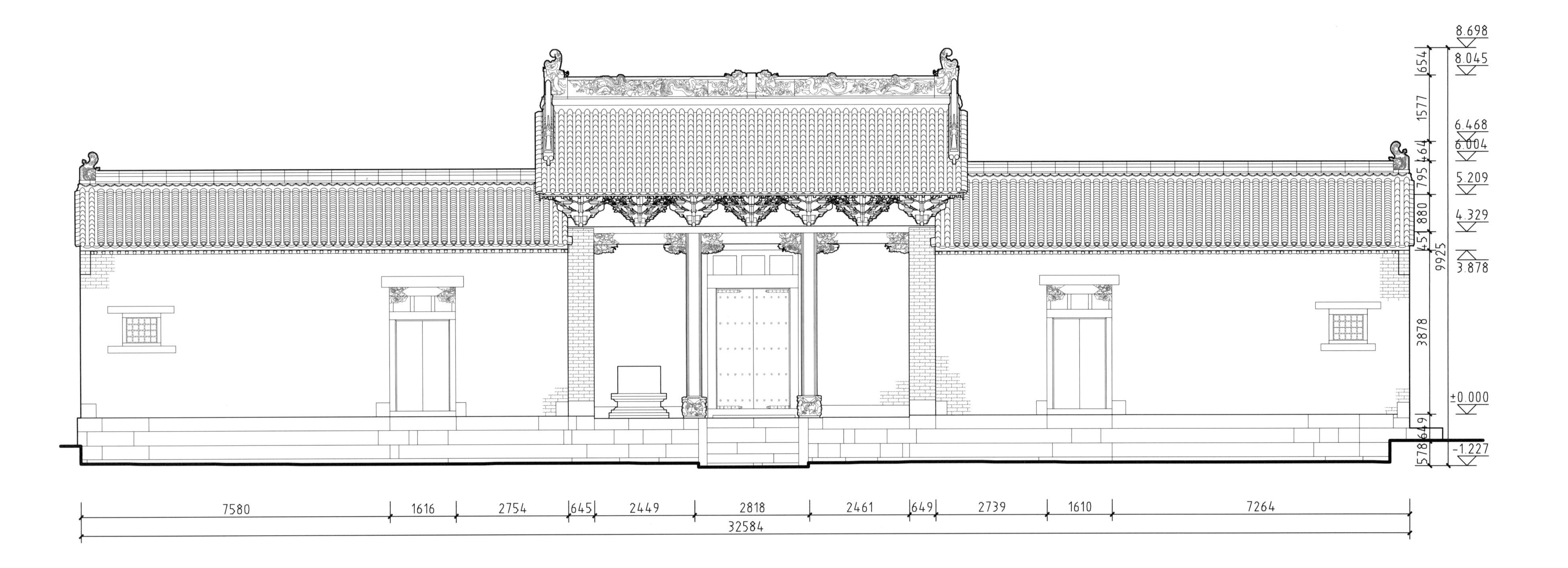

西李门二仙庙山门正立面图
Front elevation of Xilimen Erxian Temple's *shanmen*

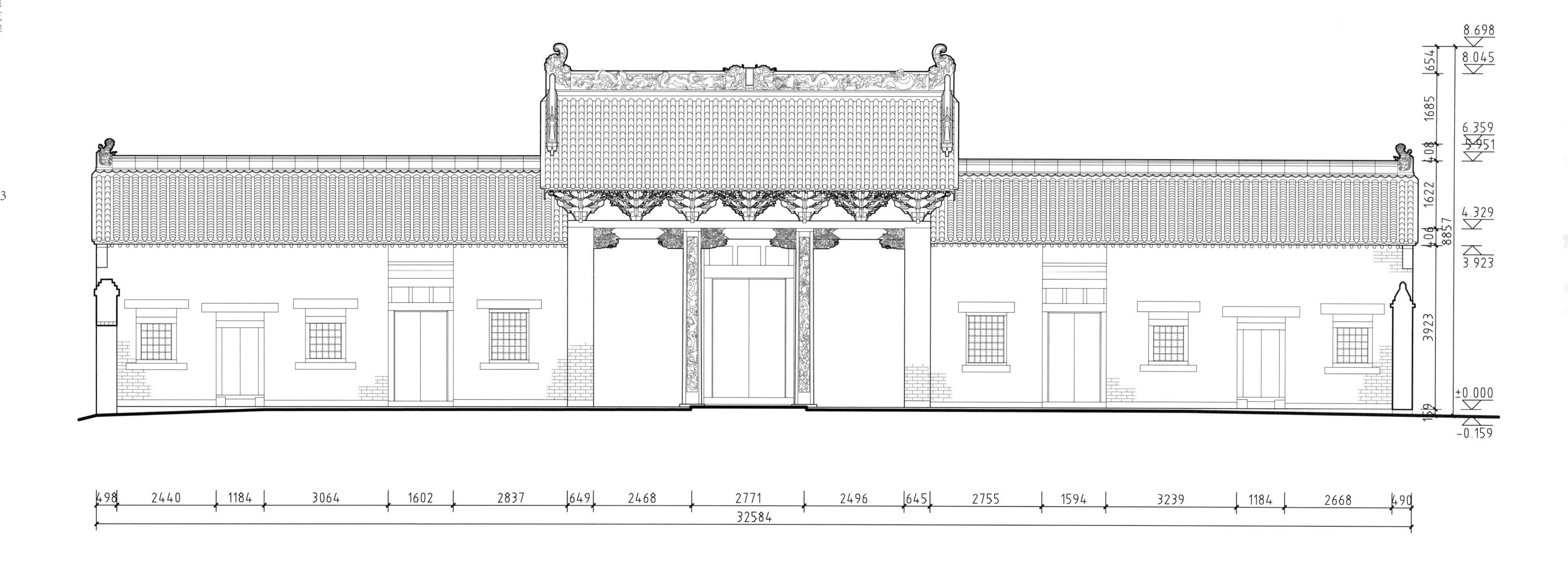

西李门二仙庙山门背立面图

Rear elevation of Xilimen Erxian Temple's *shanmen*

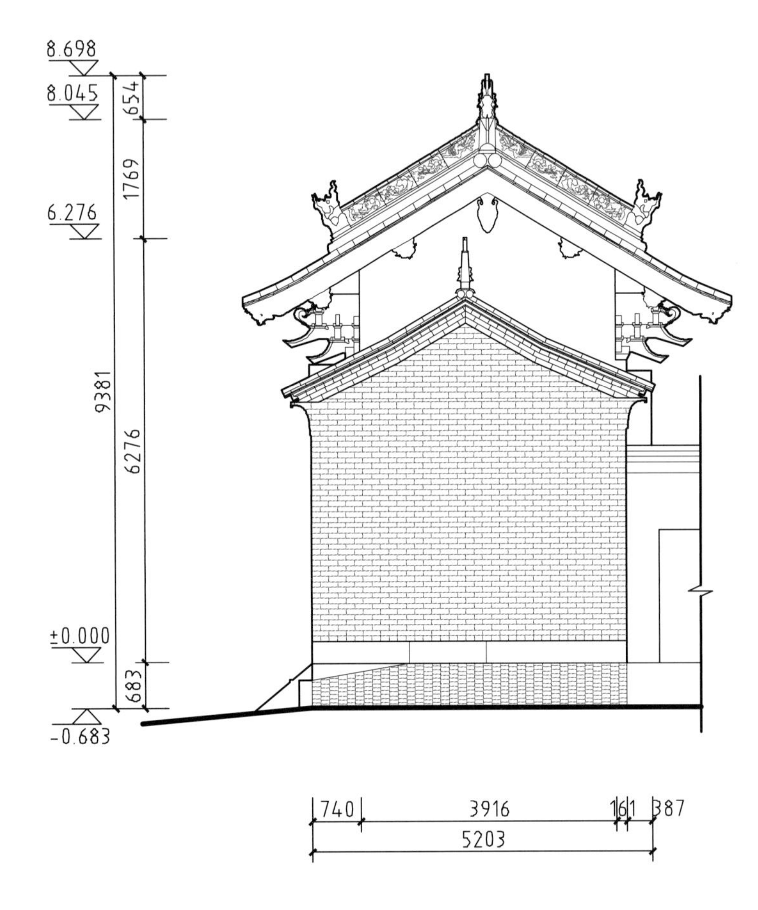

西李门二仙庙山门侧立面图
Side elevation of Xilimen Erxian Temple's *shanmen*

0 1 2.5m

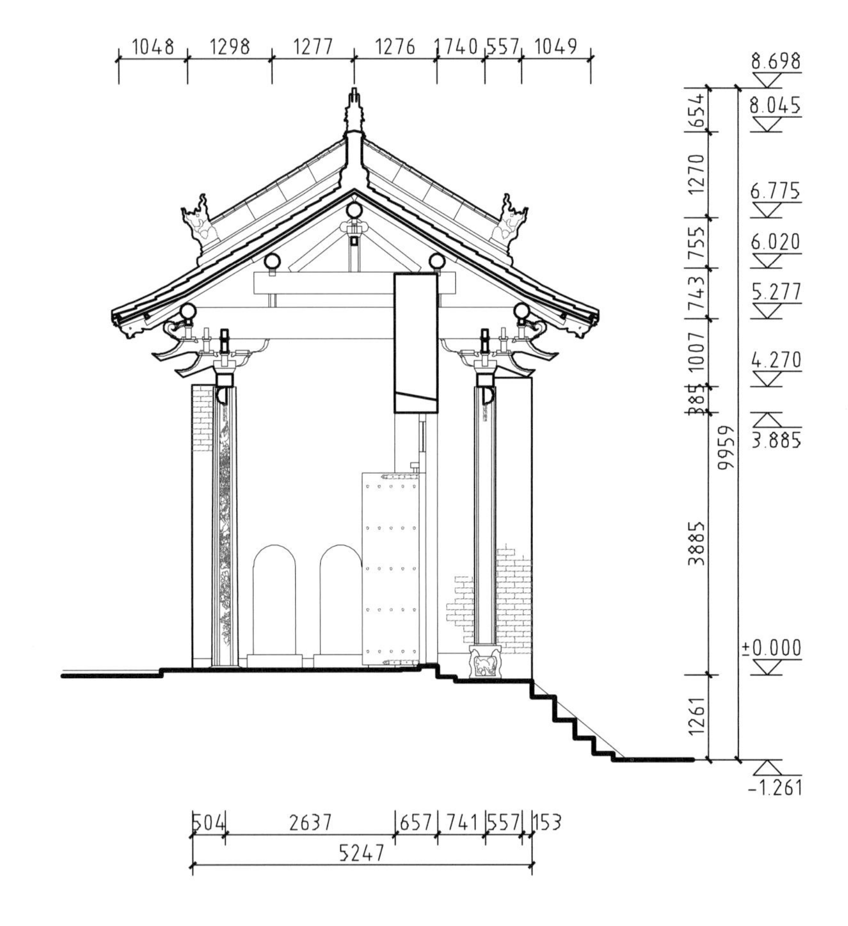

西李门二仙庙山门 A-A 剖面图
Cross-section A-A of Xilimen Erxian Temple's *shanmen*

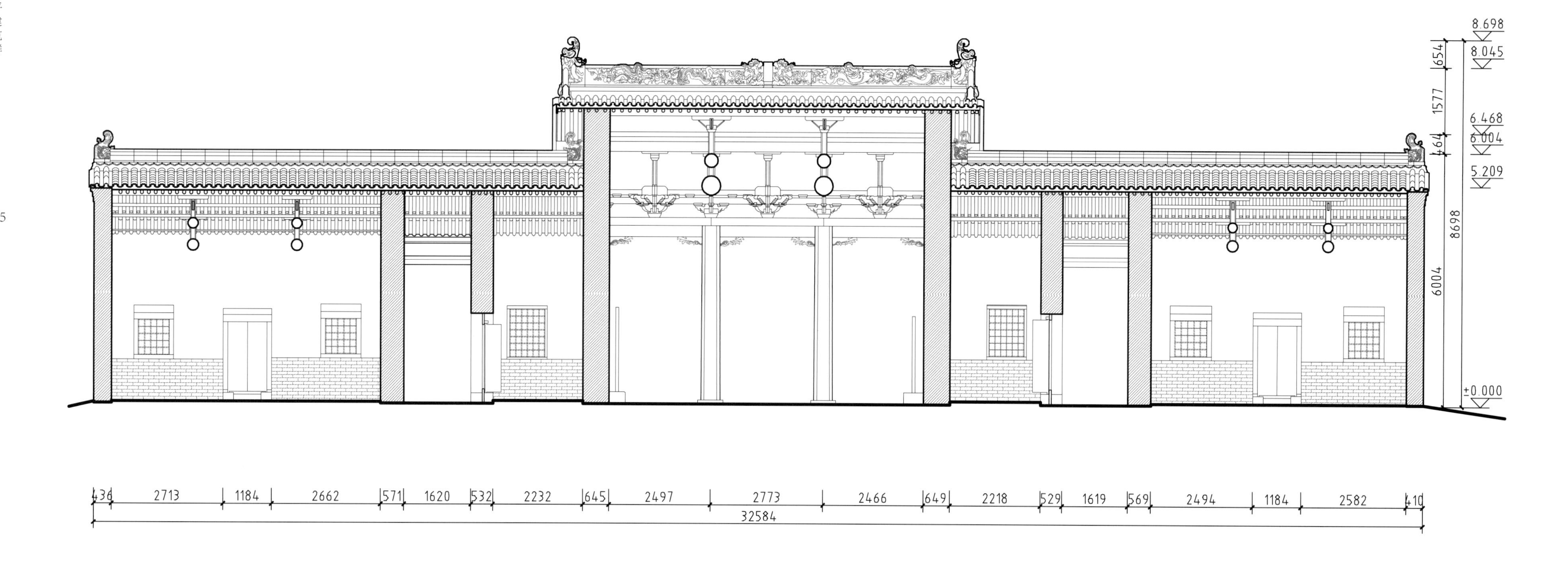

西李门二仙庙山门 B-B 剖面图

Longitudinal section B-B of Xilimen Erxian Temple's *shanmen*

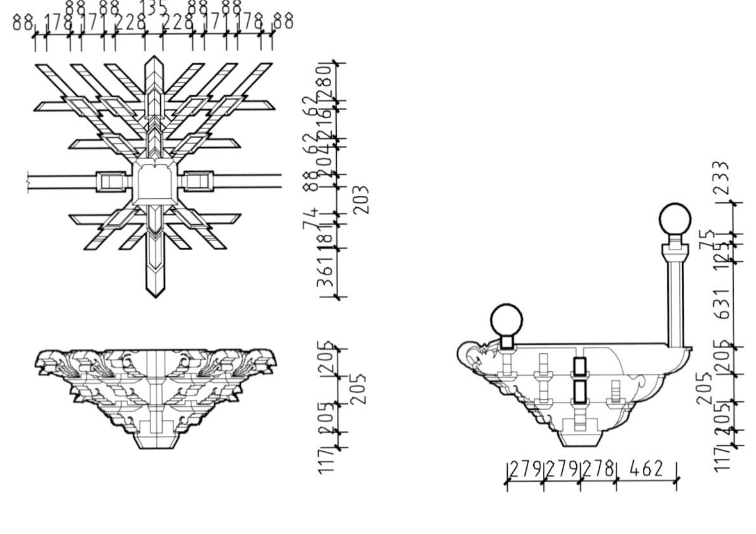

平身科大样

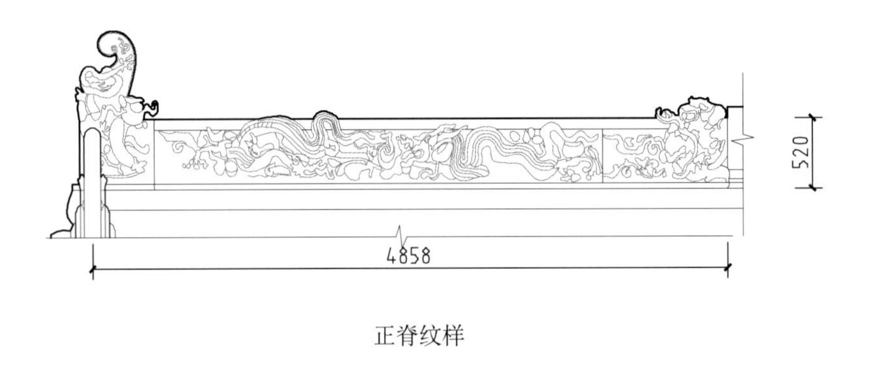

正脊纹样

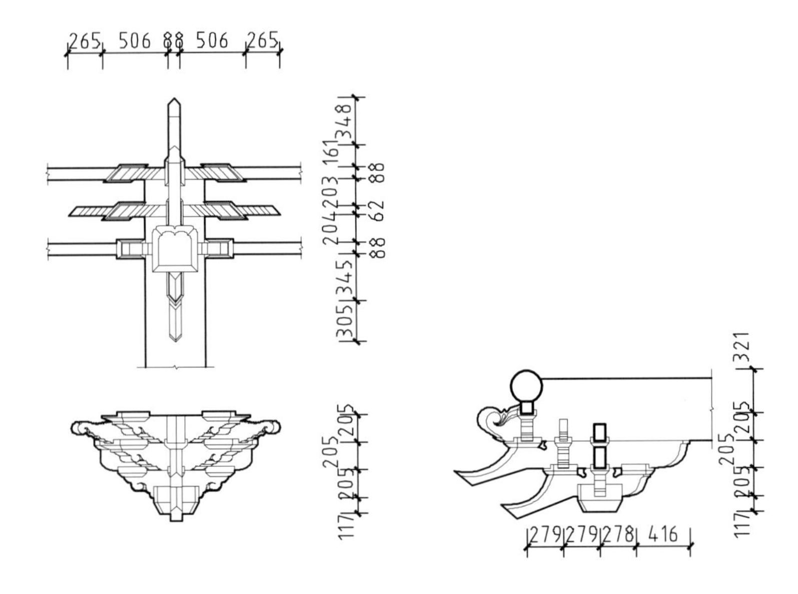

柱头科大样

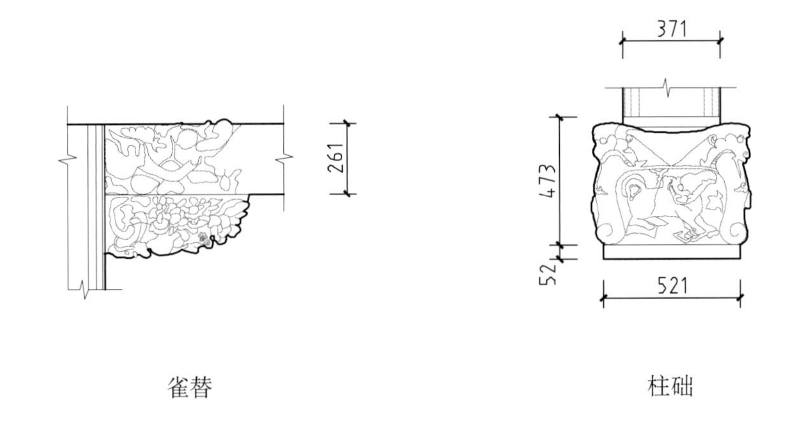

雀替

柱础

西李门二仙庙山门构件大样图

Structural component of Xilimen Erxian Temple's *shanmen*

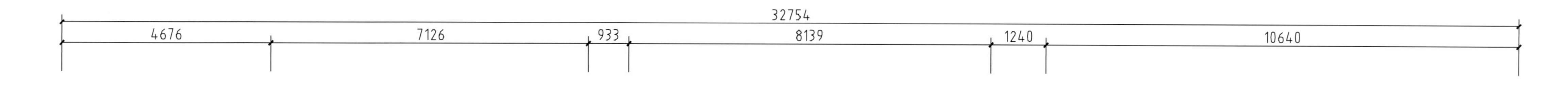

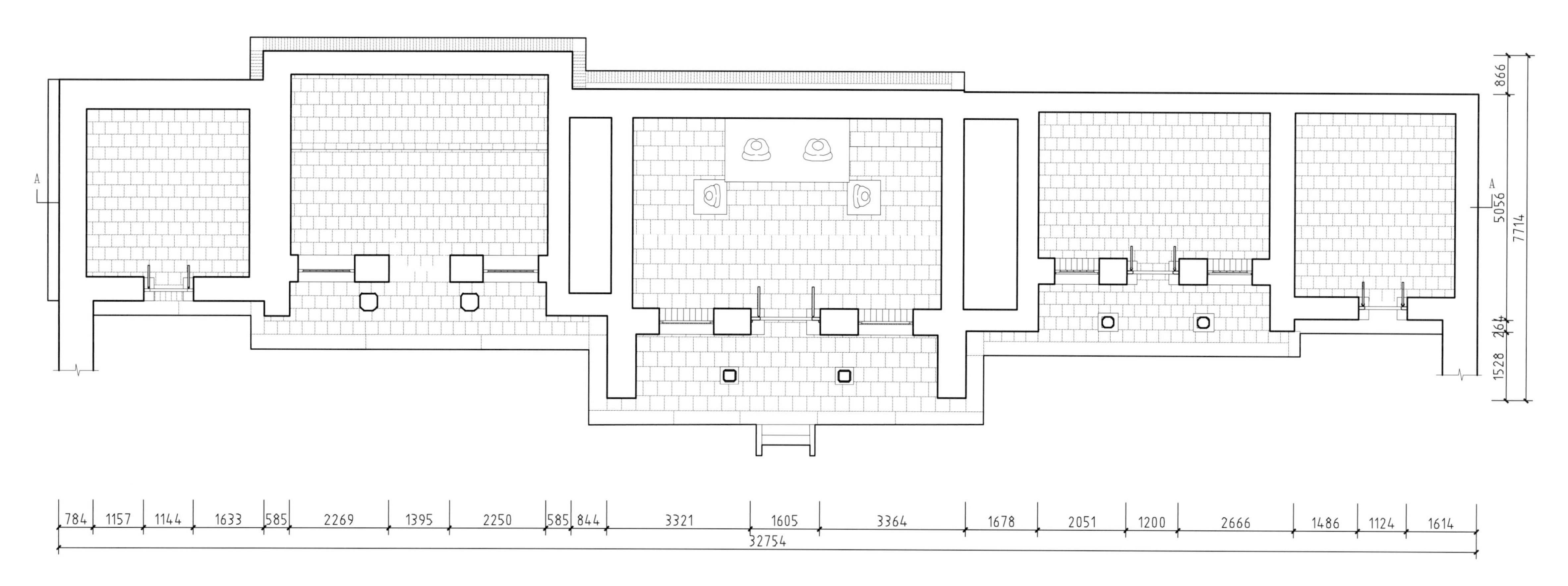

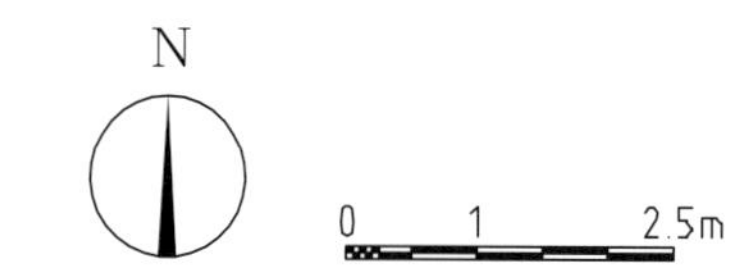

西李门二仙庙后殿平面图
Plan of Xilimen Erxian Temple's rear hall

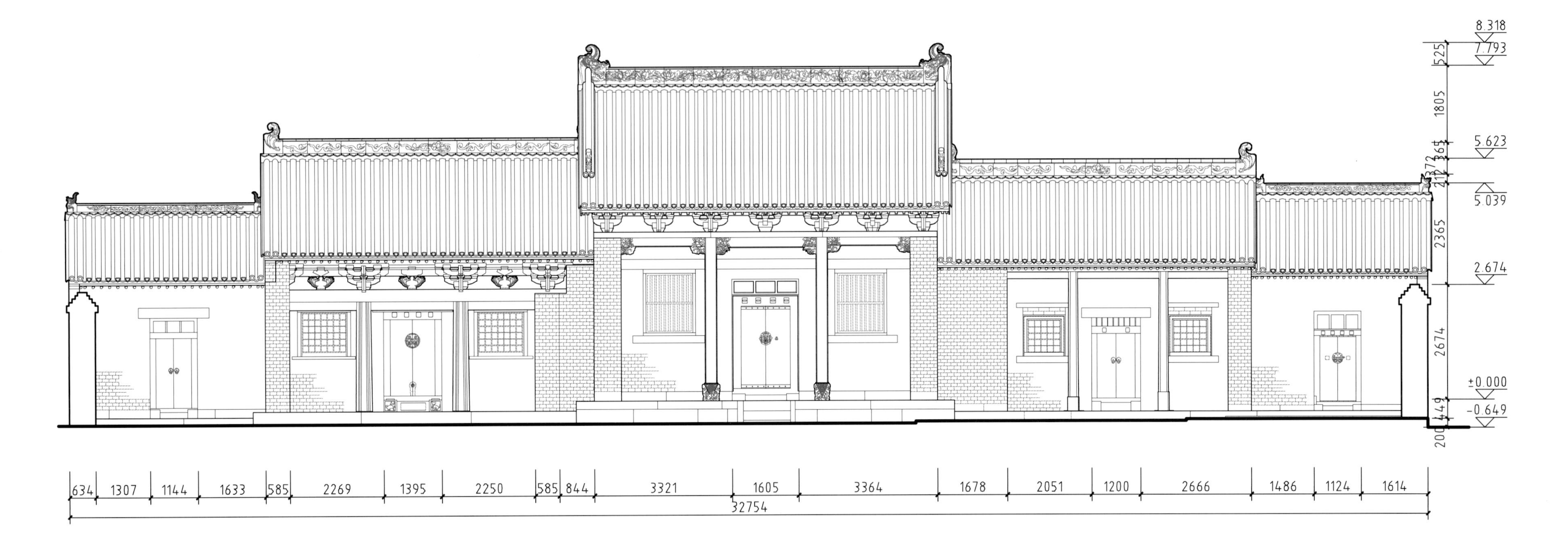

西李门二仙庙后殿正立面图

Front elevation of Xilimen Erxian Temple's rear hall

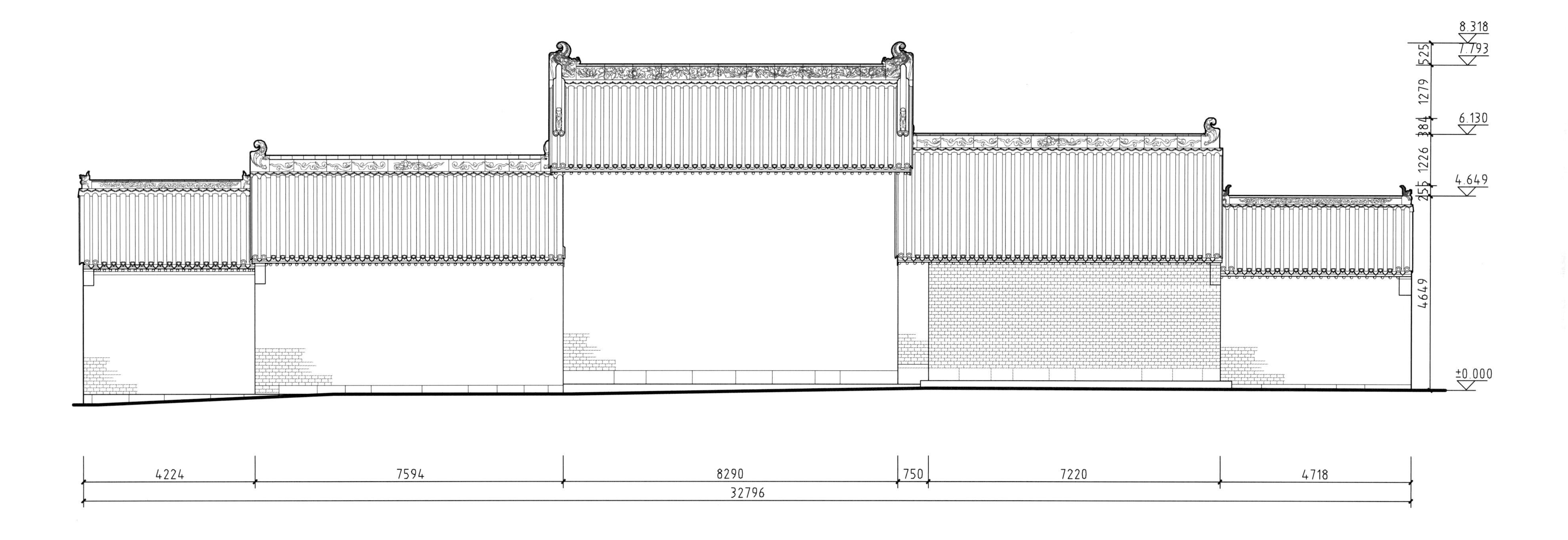

西李门二仙庙后殿背立面图

Rear elevation of Xilimen Erxian Temple's rear hall

0 1 2.5m

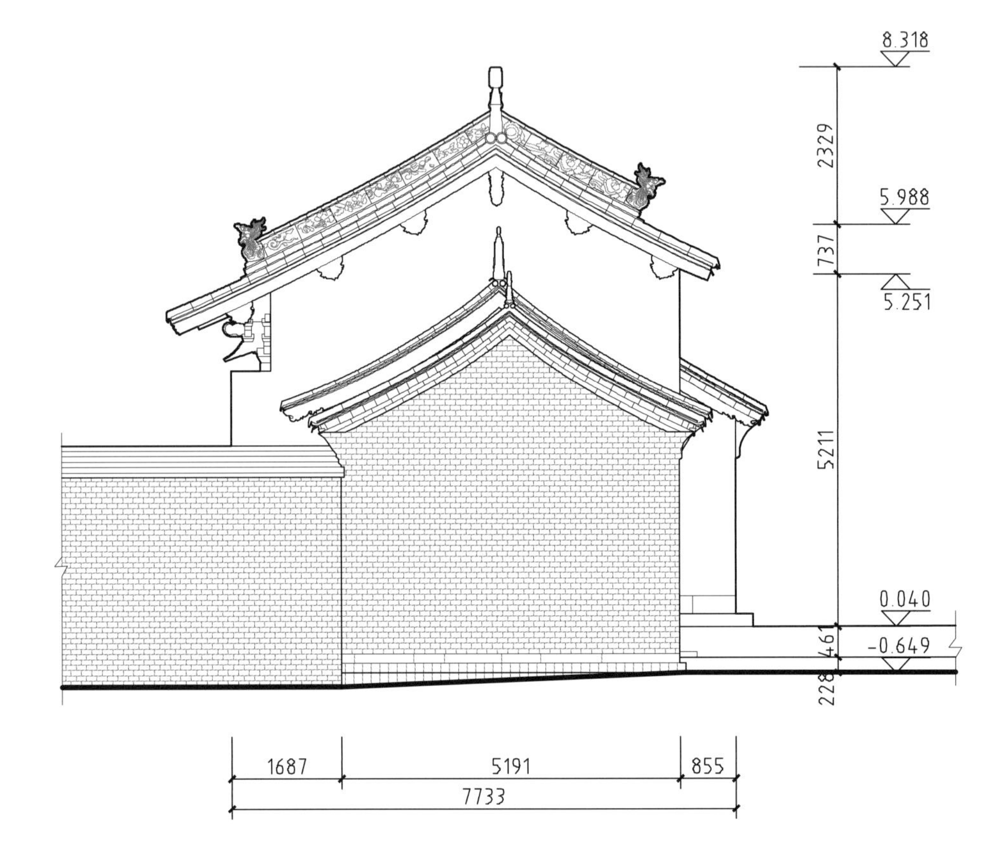

西李门二仙庙后殿东立面图
East elevation of Xilimen Erxian Temple's rear hall

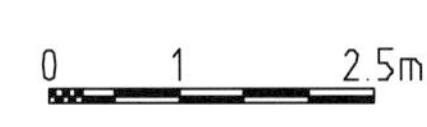

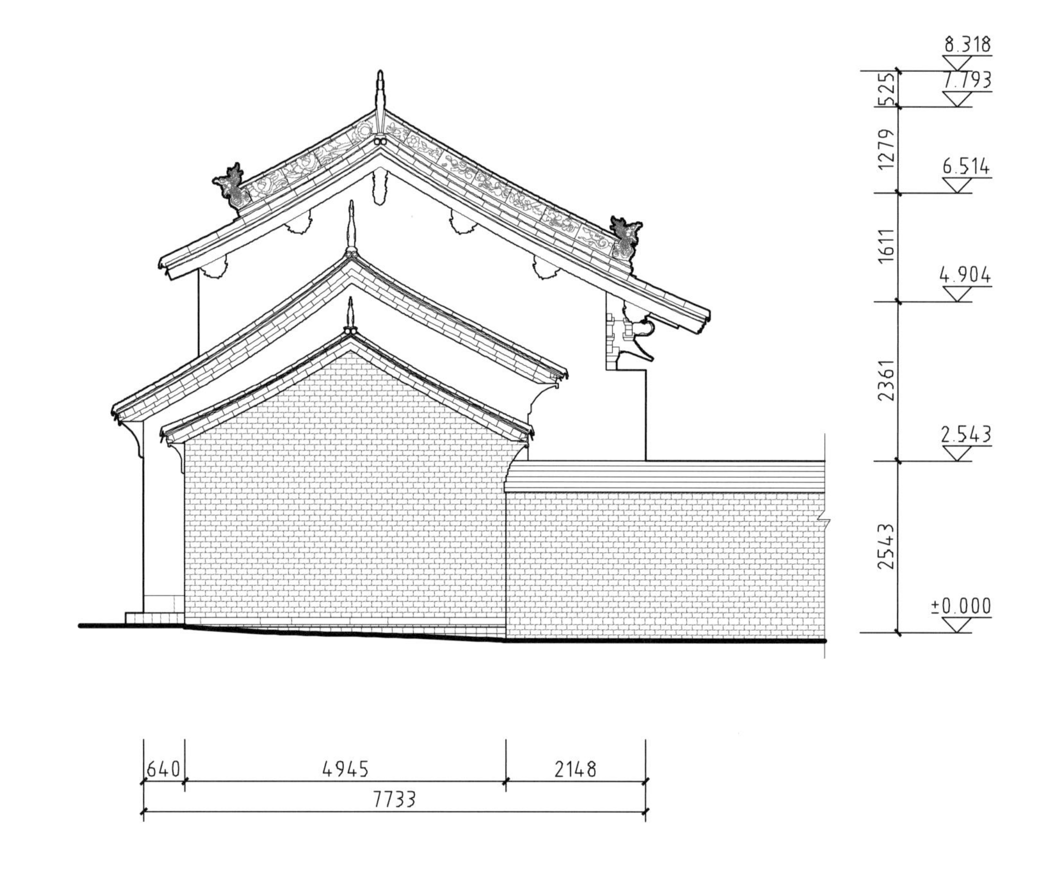

西李门二仙庙后殿西立面图
West elevation of Xilimen Erxian Temple's rear hall

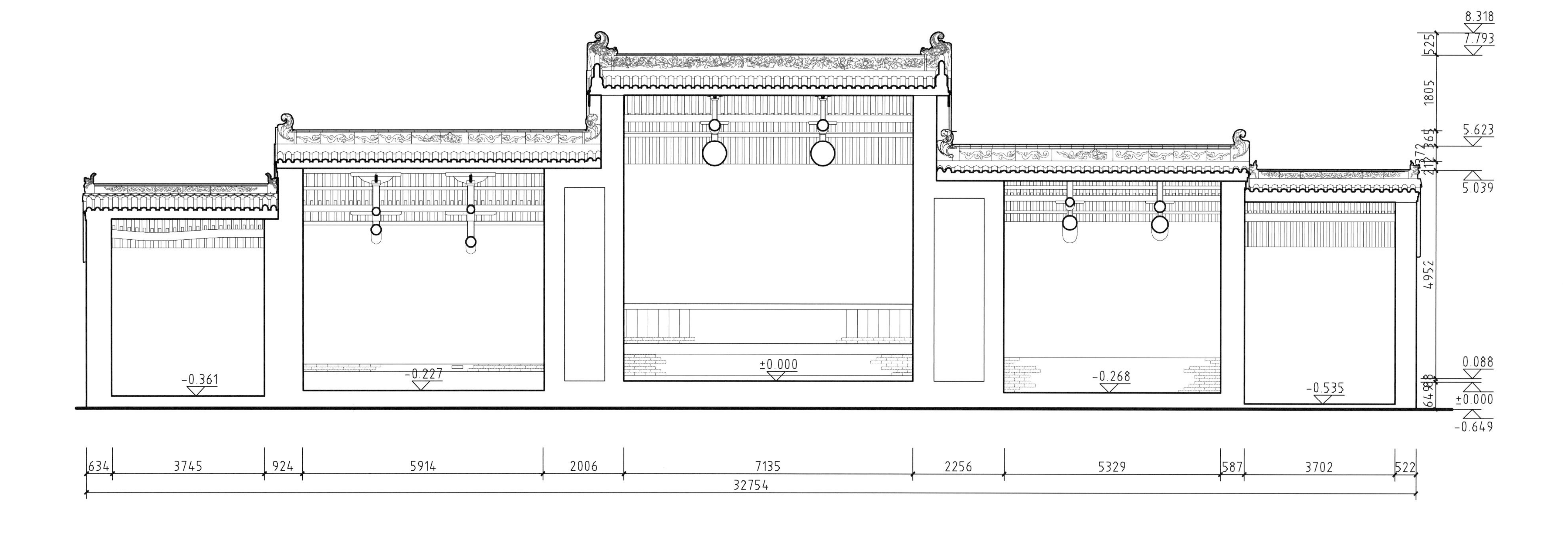

西李门二仙庙后殿 A-A 剖面图

Section A-A of Xilimen Erxian Temple's rear hall

参与测绘及相关工作的人员名单

一、崇明寺测绘人员

指导教师：贺从容　何文轩　徐　腾　黄文镐

测绘学生：江昊懋　张泽菲　丁月文　黄海璐　尚月泰　谢　娜　母卓尔
王贺楠　黎　桑　王宏升　程冰玉　于雅馨　欧阳扬　甘　草
吴　婧　冉　展　王鼎禄　李　卓

二、游仙寺测绘人员

指导教师：刘　畅　贾　珺　李沁园　蒋　哲　王曦晨　姜　铮　赵萨日娜

测绘学生：田　力　杨双琳　许通达　尤欣然　沈一琛　张　宇　许晓佳
毛俊松　张承禹　李　睿　林　璐　冉紫愚　周颖玥　曾彦玥
姚　渊　刘云松　张道琼　熊天翼　周朱盟　吕亦逊　葛思航
梁妍璐　吴崇博　卓睿珍　陈知雨　董青青　丁小涵　龚泽惠
朱锐祺　王希冉　丁文颢　熊鑫昌　达　莎　吴昭旼

三、开化寺测绘人员

指导教师：李路珂　蒋雨彤　贾　珺　张亦驰　邓阳雪　刘梦雨　李旻昊

测绘学生：周川源　杨子瑄　丁　宁　陈曦雨　黄致昊　刘诗宇　文　雯
谢恬怡　武哲睿　刘圆方　吴承霖　唐思齐　刘　田　刘倩君
付之航　苏天宇　姚　宇　胡曦阳　王佳怡　高浩歌　黎继明
唐波晗　胡毅衡

四、资圣寺测绘人员

指导教师：刘　畅　青　锋　李沁园　徐　扬　辛惠园

测绘学生：张哲敏　崔光赫　高进赫　崔成进　林雨铭　庞凌波　孙　越
张嘉琦　于博柔　祁　佳　高钧怡　张恒源　常　浩　聂　聪
裴　昱　罗哲焜　张雅敬

五、西李门二仙庙测绘人员

指导教师：王贵祥 黄文镐 杨 澍 翁 帆 徐 腾

测绘学生：项轲超 王玉颖 林浓华 黄孙扬 彭 鹏 吴濯杭
周 桐 刘 通 钱漪沅 肖玉婷 龚怡清 张昊天
唐 博 温从爽 何文轩 马志桐 孙仕轩 陈爽云
李天颖 连 璐

六、图纸整理及相关工作

图纸统筹：李 菁

图纸整理：杨 博 唐恒鲁 单梦林 买琳琳 胡竞芙 姜 铮
赵寿堂

英文统筹：［奥］荷雅丽

英文翻译：［奥］荷雅丽 Michael Norton 周彦邦

Name List of Participants Involved in Surveying and Related Works

1. Surveying and Mapping of Chongming Monastery

Supervising Instructor: HE Congrong, HE Wenxuan, XU Teng, HUANG Wenhao
Team Members: JIANG Haomao, ZHANG Zefei, DING Yuewen, HUANG Hailu, SHANG Yuetai, XIE Na, MU Zhuoer, WANG Henan, LI Sang, WANG Hongsheng, CHENG Bingyu, YU Yaxin, OUYANG Yang, GAN Cao, WU Jing, RAN Zhan, WANG Dinglu, LI Zhuo

2. Surveying and Mapping of Youxian Monastery

Supervising Instructor: LIU Chang, JIA Jun, LI Qinyuan, JIANG Zhe, WANG Xichen, JIANG Zhen, ZHAO Sarina
Team Members: TIAN Li, YANG Shuanglin, XU Tongda, YOU Xinran, SHEN Yichen, ZHANG Yu, XU Xiaojia, MAO Junsong, ZHANG Chengyu, LI Rui, LIN Lu, RAN Ziyu, ZHOU Yingyue, ZENG Yanyue, YAO Yuan, LIU Yunsong, ZHANG Daoqiong, XIONG Tianyi, ZHOU Zhumeng, LU Yixun, GE Sihang, LIANG Yanlu, WU Chongbo, ZHUO Ruizhen, CHEN Zhiyu, DONG Qingqing, DING Xiaohan, GONG Zehui, ZHU Ruiqi, WANG Xiran, DING Wenhao, XIONG Xinchang, DA Sha, WU Zhaowen

3. Surveying and Mapping of Kaihua Monastery

Supervising Instructor: LI Luke, JIANG Yutong, JIA Jun, ZHANG Yichi, DENG Yangxue, LIU Mengyu, LI Minhao
Team Members: ZHOU Chuanyuan, YANG Zixuan, DING Ning, CHEN Xiyu, HUANG Zhihao, LIU Shiyu, WEN Wen, XIE Tianyi, WU Zherui, LIU Yuanfang, WU Chenglin, Tang Siqi, LIU Tian, LIU Qianjun, FU Zhihang, SU Tianyu, YAO Yu, HU Xiyang, WANG Jiayi, GAO Haoge, LI Jiming, TANG Bohan, Hu Yiheng

4. Surveying and Mapping of Zisheng Monastery

Supervising Instructor: LIU Chang, QING Feng, LI Qinyuan, XU Yang, XIN Huiyuan
Team Members: ZHANG Zhemin, CUI Guanghe, GAO Jinhe, CUI Chengjin, LIN Yuming, PANG Lingbo, SUN Yue, ZHANG Jiaqi, YU Borou,QI Jia, GAO Junyi, ZHANG Hengyuan, CHANG Hao, NIE Cong, PEI Yu, LUO Zhekun, ZHANG Yajing

5. Surveying and Mapping of Xilimen Erxian Temple

Supervising Instructor: WANG Guixiang, HUANG Wenhao,YANG Shu, WENG Fan, XU Teng
Team Members: XIANG Kechao, WANG Yuying, LIN Nonghua, HUANG Sunyang, PENG Peng, WU Quhang, ZHOU Tong, LIU Tong, QIAN Yiyuan, XIAO Yuting, GONG Yiqing, ZHANG Haotian, TANG Bo, WEN Congshuang, HE Wenxuan, MA Zhitong, SUN Shixuan, CHEN Shuangyun, LI Tianying, LIAN Lu

6. Editor of Drawings and Related Works

Drawings Arrangement: LI Jing
Drawings Editor: YANG Bo, TANG Henglu, SHAN Menglin, MAI Linlin, HU Jingfu, JIANG Zheng, ZHAO Shoutang
Translator in Chief: Alexandra Harrer
Translation Members: Alexandra Harrer, Michael Norton, CHOU Yen Pang

图书在版编目（CIP）数据

高平建筑群＝ARCHITECTURE COMPLEX OF GAOPING / 清华大学建筑学院编写；刘畅，王贵祥，廖慧农主编. — 北京：中国建筑工业出版社，2019.12
（中国古建筑测绘大系 . 宗教建筑）
ISBN 978-7-112-24557-4

Ⅰ.①高… Ⅱ.①清… ②刘… ③王… ④廖… Ⅲ. ①宗教建筑—建筑艺术—高平—图集 Ⅳ.①TU-885

中国版本图书馆CIP数据核字（2019）第286221号

丛书策划 / 王莉慧
责任编辑 / 李 鸽 陈海娇
英文审稿 / [奥] 荷雅丽（Alexandra Harrer）
书籍设计 / 付金红
责任校对 / 张惠雯

中国古建筑测绘大系 · 宗教建筑
高平建筑群
清华大学建筑学院 编写
刘畅 王贵祥 廖慧农 主编

Traditional Chinese Architecture Surveying and Mapping Series: Religious Architecture
ARCHITECTURE COMPLEX OF GAOPING
Compiled by School of Architecture, Tsinghua University
Edited by LIU Chang, WANG Guixiang, LIAO Huinong

*

中国建筑工业出版社出版、发行（北京海淀三里河路 9 号）
各地新华书店、建筑书店经销
北京方舟正佳图文设计有限公司制版
北京雅昌艺术印刷有限公司印刷

*

开本：787 毫米 × 1092 毫米 横 1/8 印张：18 字数：474 千字
2021 年 6 月第一版 2021 年 6 月第一次印刷
定价：**178.00** 元
ISBN 978-7-112-24557-4
（35102）